T0192335

Paddy Soil Science

Kazutake Kyuma

Kyoto University Press

First published in 2004 jointly by:

Kyoto University Press
Kyodai Kaikan
15-9 Yoshida Kawara-cho
Sakyo-ku, Kyoto 606-8305, Japan
Telephone: +81-75-761-6182
Fax: +81-75-761-6190
Email: sales@kyoto-up.gr.jp
Web: http://www.kyoto-up.gr.jp

Trans Pacific Press
PO Box 120, Rosanna, Melbourne
Victoria 3084, Australia
Telephone: +61 3 9459 3021
Fax: +61 3 9457 5923
Email: info@transpacificpress.com
Web: http://www.transpacificpress.com

Copyright © Kyoto University Press and Trans Pacific Press 2004

Printed in Melbourne by BPA Print Group

Distributors

Australia
Bushbooks
PO Box 1958, Gosford, NSW 2250
Telephone: (02) 4323-3274
Fax: (02) 4323-3223
Email: bushbook@ozemail.com.au

UK and Europe
Asian Studies Book Services
Fransweg 55B
3921 DE Elst
Utrecht
The Netherlands
Telephone: +31 318 470 030
Fax: +31 318 470 073
Email: info@asianstudiesbooks.com

USA and Canada
International Specialized Book
Services (ISBS)
5824 N. E. Hassalo Street
Portland, Oregon 97213-3644
USA
Telephone: (800) 944-6190
Fax: (503) 280-8832
Email: orders@isbs.com
Web: http://www.isbs.com

All rights reserved. No production of any part of this book may take place without the written permission of Kyoto University Press or Trans Pacific Press.

ISBN 978-1-920901-00-4

National Library of Australia Cataloging in Publication Data

Kyuma, Kazutake.
Paddy soil science.
Bibliography.

Includes index.
ISBN 978-1-920901-00-4.

1. Soil science. 2. Rice - Soils. 3. Soils - Asia. I.
Title.

631.4

Contents

List of Figures

List of Tables

Foreword

Many soil science textbooks have been written and published in English and so there should not be any difficulty for students to choose a text matched to their desired level of knowledge, from elementary to advanced. However, these textbooks, do not normally have detailed descriptions of paddy soils or those soils that undergo seasonal submergence and drainage for rice cultivation.

In Monsoon Asia, students who are studying agricultural science should have a basic knowledge of the nature and properties of paddy soils, as this is the medium for production of their most important food staple. Unfortunately, there has so far been no detailed textbook on paddy soils written in English.

The present volume is an attempt to fill this serious gap. The author had worked on paddy soils in tropical Asia, as well as in Japan, for many years until his retirement from Kyoto University, and later he had opportunities to give lectures on 'paddy soil science' at Bogor Agricultural University in Indonesia and Kasetsart University in Thailand. Last year he began reworking his lecture notes to make them more up-to-date and more comprehensive, relying mainly on material published in Japan.

The author borrowed information from the text Suiden-Dojyo-Gaku (Paddy Soil Science) that was edited by Prof. Keizaburo Kawaguchi and published in 1978 by Kodansha of Tokyo. This book integrated the results of research conducted in Japan from the late Meiji era in the early 20th century, when scientific studies on paddy soils in relation to rice began. Thirty-six members of the Japanese Society of Soil Science and Plant Nutrition contributed to this substantial book. The author owes a great deal to those contributors for the contents of the present volume, particularly of its Chapters 3, 5, 8 and 9, and wishes to express his heartfelt thanks to all of them. Another important source of information on paddy soil fertility was the recent review papers that appeared in Volume 67 (1996) of the *Japanese Journal of Soil Science and Plant Nutrition*: 'Nitrogen' by Dr. K. Toriyama, 'Phosphorus' by Dr. M. Nanzyo, and 'Silicic acid' by Dr. H. Sumida. The information from these articles that is relevant to the aims of this book is introduced in Chapters 8 and 9. Also, the author learnt about recent advances in nitrogen fixation from Dr. T. Isoi and obtained information on soil enzymes and weeds from Dr. S. Kanazawa and Dr. S. Miyagawa, respectively. The author wishes to acknowledge, with many thanks, the contributions of these scientists to this book.

The author acknowledges with great appreciation the kindness with which the following scientists took the trouble of reviewing various manuscripts for this book: Dr. K. Minami, Dr. K. Yagi, and Dr. H. Kato for Chapters 4 and 6, Dr. K. Toriyama and Dr. T. Isoi for Chapter 8, and Dr. H. Sumida for Chapter 9.

The author is deeply indebted to Dr. H. Eswaran of the Unites States Department of Agriculture. He not only reviewed both the contents and English of all the chapters, but also made several suggestions that improved the usefulness of the book.

The author made alterations and corrections according to the suggestions received from his colleagues, but any errors and misconceptions still in the book are solely his responsibility.

As stated earlier, Bogor Agricultural University, particularly its Center for Wetland Studies, gave the author a Guest Professorship in September 1998. The author gratefully acknowledges Prof. Supiandi Sabiham, then Director of the Center, and Dr. Widiatmaka, DAA of the Department of Soil Science, for making his three week stay in Bogor a very pleasant experience. Furthermore, the Center helped compile the lecture notes. The author expresses his most sincere thanks to the Center and its staff.

The Graduate School of Kasetsart University, with Prof. Tasnee Attanandana as Dean, offered the author the privilege of the visiting professorship in 2001-2002. The author gave a course on paddy soil science to graduate students at the Department of Soil Science and worked further on his lecture notes to bring them to the present form. The author is most grateful to Prof. Tasnee Attanandana for her keen interest in and generous understanding of his work.

The Japan Society for the Promotion of Sciences (JSPS) allocated the 2003 Grant-in-Aid for Publication of Scientific Research Results (No. 155307) of KAKENHI (Grant-in-Aid for Scientific Research) to the publication of this book, through Kyoto University Press and Trans Pacific Press. The author wishes to express his most sincere thanks for this financial assistance, without which this book would not have been published.

May 15th, 2003.
KYUMA Kazutake

Chapter 1: Introduction

1.1 The need for paddy soil science

The English term, 'paddy,' means either a rice plant, unhusked rice grains, or a rice field. It is derived from the Malay word *padi* meaning rice, irrespective of grouping such as hill *padi* or *padi sawah*. Thus, in its original meaning, paddy does not necessarily signify anything related to the practice of rice cultivation under submerged or waterlogged conditions. However, since by far the large majority of rice is cultivated under submerged conditions in Asia, the word paddy today implicitly means aquatic rice or lowland rice. When one wishes to specify rice grown under upland conditions, the term 'upland rice' or 'hill paddy' is used to distinguish the crop from ordinary aquatic paddy or paddy *sawah*. Paddy in this monograph refers to aquatic rice.

What is a paddy soil? It is a soil used or potentially usable for growing aquatic rice. In this definition, any soil occurring in a climatic zone with a temperature regime that is suited to growing one crop of rice a year is called paddy soil, if sufficient water is available to submerge the soil for the necessary length of time needed by the crop. Thus, temperature and water are two major constraints. Consequently, as will be elaborated in the next chapter, the distribution of paddy lands is confined to, or concentrated in, those parts of the world where water is available in abundance.

As stated above, the term 'paddy soil' is related directly to land use, but not to any particular types of soil in the pedological sense. Dudal (1958) listed the following soil groups as those frequently used for growing paddy rice: alluvial soils, gray hydromorphic soils (or low humic gley soils), grumusols, latosols, andosols, regosols, red-yellow podzolic soils, planosols, and gray brown podzolic soils. Eswaran *et al.* (2001) also listed all the soil orders of Soil Taxonomy (1999) except Gelisols as those soils used for paddy rice cultivation, but they also indicated that Inceptisols and Ultisols are the most frequently used.

Ambiguous as it is, the term 'paddy soil' is still preferable to such terms as 'rice soil' and 'submerged soil'. 'Rice soil' implies both upland and lowland rice-growing soils, whilst the term 'paddy soil' is customarily used

1

to signify lowland rice soils (or *sawah* soils in the Malay language). 'Submerged soil' means the state of soil being under water. Of course, submergence, or waterlogging, is the key factor that brings about most of the unique properties and processes of paddy soil, but some of the characteristics of paddy soil are the products of a cyclic process of alternating submergence and drying (or draining) and not just the result of submergence.

Why, then, are we so interested in paddy soils? There are two main reasons. First, rice, the product of paddy soils, is an important cereal crop in global agriculture. Table 1.1 (FAO, 2001) shows that rice is second, after wheat, in terms of the harvested area of cereal crops, although the difference is still considerable. Maize is becoming increasingly important both in area and production, but it is mainly cultivated for animal feed and not for immediate human consumption. Thus, most noteworthy is the fact that the rice production figure has surpassed that of wheat and maize, forming about 30% of the world's total cereal output

The second reason is that paddy lands are concentrated in Monsoon Asia, with a little more than 90% of the world rice growing area occurring in this region. This is immediately clear from Figure 1.1, which shows the global distribution of rice lands. In the countries of Monsoon Asia, frequently more than half the total arable land is devoted to aquatic rice cultivation. Thus, paddy soils are one of the most important groups of soils for soil scientists living in Monsoon Asia.

These observations provide important justifications for establishing a special branch of soil science called 'paddy soil science' to distinguish it from classical 'soil science'. First, paddy soils are now one of the world's most important media for food production. Second, in spite of their

Table 1.1 Importance of rice in the world food production

Crop	Area harvested ha	Production t	Yield t ha^{-1}
Wheat	213,600,203	576,317,042	2.70
Rice	153,765,832	598,851,733	3.89
Maize	139,681,860	590,790,877	4.23
Barley	57,190,141	131,990,121	2.31
Sorghum	42,070,697	58,500,218	1.39
Millet	35,974,793	27,254,750	0.76
Cereals	675,405,416	2,049,414,990	3.03

(Source: FAOSTAT, 2001)

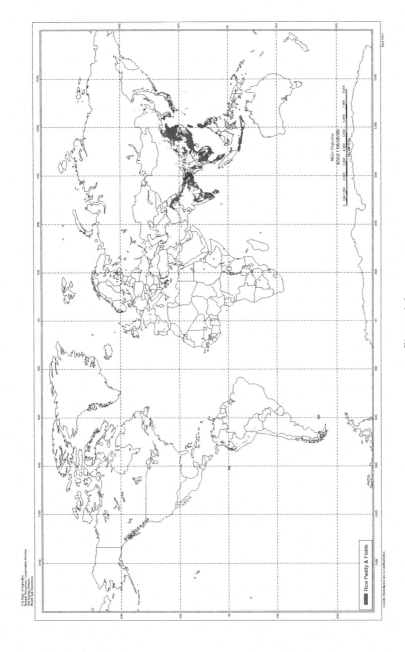

Figure 1.1

Map of the distribution of world paddy lands (Courtesy Dr. H. Eswaran, USDA)

importance, ordinary soil science textbooks do not devote adequate attention to paddy soil properties, processes, and genesis, partly because of their localized occurrence in Monsoon Asia and their unusual method of management, that involves submergence. This latter point has to be stressed further as a third justification. Managing paddy soils under water is entirely different from the management of other soils used for upland crops, and produces important differences, particularly in chemical and biochemical or microbiological processes. The cyclic change in micro-environmental conditions expresses properties not encountered in other soils and differentiates paddy soils from most other soil systems.

Although many studies have been conducted on paddy soils in different parts of the world, the amount of experience and data that has been accumulated in Japan is exceptional and exceeds similar information available in most other countries. Many of these studies however, are written in the Japanese language, and so are inaccessible to the international soil science community. This was another important motivation for the author to summarizing all available data in a systematic way as paddy soil science—so that anyone can use the experiences and views of Japanese farmers and researchers.

1.2 A brief historical review

The first notion of the specific behavior of paddy soils came from Daikuhara and Imazeki (1907) in relation to the behavior of nitrate, that is, denitrification and its acceleration by the application of organic substrates. Gillespie (1920) and others reported a lowering of the redox potential (Eh) and its enhancement with an application of organic substances to submerged soils. However, these studies only paid attention to an individual phenomenon and still lacked the comprehensive understanding that these processes were interlinked, with the redox system serving as the common denominator.

In 1936, Osugi and Takusima published a paper entitled '*A few differences between the soils under paddy field conditions and those under upland field conditions*', and reported a reduction of various substances under submergence, an increase of exchange acidity due to the formation of ferrous ions, and an increase in the solubility of soil phosphates in paddy soils because of the reduction process. As suggested by the title, this may be the first work that clearly noted the specificity of paddy soils compared with upland soils.

In 1937, Osugi and Morita published a paper entitled *'On the investigation of soils in paddy and dry field conditions,'* and reported the occurrence of an illuvial horizon of iron oxides as a specific morphological feature of paddy soils resulting from the mobilization of active iron within the profile. As explained in Chapter 7, this is one of the main items of evidence of the redoximorphic features found in paddy soils.

Also in 1937, Shioiri and Aomine reported a denitrification phenomenon that occurred in the surface horizon of paddy soils because of differentiation of a thin oxidized layer at the soil/water interface after several weeks of submergence.

In 1938, Osugi and Kawaguchi confirmed the formation of hydrogen sulfide in paddy soils as a result of sulfate reduction, and this work opened up the way to elucidation of the *akiochi* phenomenon (see Chapter 9). Shioiri and his colleagues finally clarified the cause of *akiochi* and related phenomena as being the degraded nature of soil materials, including the deficiencies of readily reducible iron oxides and nutrient cations (Shioiri, 1944).

These and other early findings contributed to many special studies on paddy soils and stimulated the development of innovative technologies to evaluate *in vitro* processes operating under submerged conditions. Such studies, and specifically the details that were to be considered, are not replicated in many other groups of soils or employed to describe soil conditions. This opened up a path to a new field of research on paddy soils. The objective of this monograph is to integrate and summarize all available relevant information. The author hopes that paddy soil researchers in Monsoon Asia will find this useful as they build on previous work to ensure food security, and more specifically, to guarantee the supply of rice to the people of Asia.

References

Daikuhara, G. and Imaseki, T. 1907. On the behavior of nitrate in paddy soils. *Bul. Imp. Central Agric. Expt. Sta.*, 1(2): 7–36. (Cited from Kawaguchi, K. (Ed.), 1978. *Paddy Soil Science*. Kodansha, Tokyo: In Japanese).

Dudal, R. 1958. Paddy soils. *International Rice Commission, News Letter*, 7(2): 19–27.

Eswaran, H., Moncharoen, P., Reich, P. and Padmanaban, E. 2001. Rice, land, and people: The faltering nexus in Asia. *Proc. of the 5th*

Conference of East and Southeast Asia Federation of Soil Science Societies, Krabi, Thailand, pp. 38–66. Dept. Agriculture, Bangkok, Thailand.

Food and Agriculture Organization 2001. FAOSTAT. FAO, Rome.

Gillespie, L.J. 1920. Reduction potentials of bacterial cultures and of waterlogged soils. *Soil Sci.*, 9: 199–216. (Cited from Kawaguchi, K. (Ed.), 1978. *Paddy Soil Science*. Kodansha, Tokyo: In Japanese).

Osugi, S. and Kawaguchi, K. 1938. On the reduction of sulfates in paddy field soil. *J. Sci. Soil & Manure, Japan*, 12: 453–462. (In Japanese).

Osugi, S. and Morita, S. 1937. On the investigation of soils in paddy and dry field conditions. *J. Sci. Soil & Manure, Japan*, 11: 355–68. (In Japanese).

Osugi, S. and Takusima, T. 1936. A few differences between the soils under paddy field conditions and those under upland field conditions. *Agric. & Hortic.*, 11: 1655–1660. (In Japanese).

Shioiri, M. 1944. *Root Rot and 'Akiochi' of Rice in Degraded Paddy Soil.* (In Japanese) (Cited from Kawaguchi, K. (Ed.), 1978. *Paddy Soil Science*. Kodansha, Tokyo: In Japanese).

Shioiri, M. and Aomine, S. 1937. Fate of ammoniacal nitrogen in the soil under paddy field condition. *J. Sci. Soil & Manure, Japan*, 11: 389–392 (Abstract). (In Japanese).

Soil Survey Staff 1999. *Soil Taxonomy, 2nd Ed.: A Basic System of Soil Classification for Making and Interpreting Soil Surveys*. USDA Agric. Handb. No.436, U.S. Gov't Printing Office, Washington, D.C.

Chapter 2: The Environmental Setting of Paddy Soils

2.1 The distribution of rice cultivation

Rice, wheat and maize occupy more than 75% of the world's harvested area of cereal crops and comprise more than 86% of the world's total cereal production. They are the three major cereals of global agriculture (see Table 1.1). The origin, manner of propagation, and the global distribution of these three cereals show important historical differences. Wheat originated in the Near and Middle East some 10000 years ago, and very quickly expanded to northern Africa and Europe and even to the Far East. Being a temperate crop, it is confined to high altitudes in the tropics. At an unknown period in time, maize was domesticated in Central America. It was introduced to Europe in the early 16th century, and was propagated rapidly throughout the world thereafter. Maize is now one of the most widely cultivated crops in the world, as it has been adapted to different climatic and soil conditions.

Rice is said to have originated somewhere in southeastern China, either in the middle or lower reaches of the Yangtze River or in the Yunnan-Assam region. Archaeological evidence of rice cultivation was found in Hemudu, near Hanzhou (south of Shanghai), dating back to 7000 years B.P. (before the present). More recently, it was reported that even older (7500–8500

Table 2.1   Area and production of rice in the world, Asia, Monsoon Asia, and tropical Asia (2000)

	Harvested area × 10³ ha (%)		Production × 10³ t (%)		Yield t paddy ha⁻¹
World	153,766	(100.0)	598,852	(100.0)	3.89
Asia	137,600	(89.5)	545,477	(91.0)	3.96
Monsoon Asia*	136,462	(88.7)	541,932	(90.4)	3.97
Tropical Asia**	102,582	(66.7)	331,144	(55.3)	3.23

* East, Southeast and South Asia   ** Southeast and South Asia
(Source: FAOSTAT, 2001)

years B.P.) archaeological evidence of rice cultivation was unearthed at Pengtoushan, Hunan Province, in the middle reaches of the Yangtze (Greenland, 1997). Thus, the origin of rice is probably similar to that of wheat, or some 10000 years ago. However, there is an important difference between wheat and rice and their subsequent propagation. Wheat has become an important food staple in Europe, North and South America, Australia, and parts of Africa and Asia, whereas rice cultivation has been confined to the regions adjacent to its origin, or what we term 'Monsoon Asia', even 10000 years after its initial domestication. As shown in Table 2.1, a little more than 90% of the production and almost the same percentage of the harvested area of rice occur in Monsoon Asia. There are a number of reasons, both physical and cultural, for this very uneven distribution of global rice cultivation. Some of the physical factors are elaborated below.

2.2 Monsoon Asia and its characteristic features

The term 'Monsoon Asia' has already been used, but where is its domain? What are its defining features? What has contributed to its extraordinarily high concentration of rice cultivation? In order to understand the environmental setting in which this concentration has taken place, it is necessary to examine the natural conditions of Monsoon Asia in more detail.

2.2.1 The monsoon and Monsoon Asia
The monsoon is a seasonal wind, whose prevailing direction varies with the season. A monsoon climate has seasonal winds that change their direction by almost 180° through a year. For example, most regions of South and Southeast Asia have southwesterly winds during the summer months, while northeasterly winds prevail during the winter months. Southwesterly winds blow across the Indian Ocean and are saturated with moisture, thus bringing the rainy season to most of tropical Asia. The northeasterly winds are blocked by mountain ranges in the north and northeast, so moisture does not reach the land, resulting in the dry season. In this way, monsoons also govern the rainfall pattern, exerting a major influence on human life, particularly on agriculture and food production.

Monsoon Asia encompasses East Asia, Southeast Asia and South Asia, but the extent of Monsoon East Asia is the most difficult to define. Greater parts of the Japanese Archipelago, the Korean Peninsula, and the coastal zone of China are doubtless under the influence of the monsoon, but how much of inland China should be included is rather ambiguous. A map

showing the soil regions of China (Gong Zitong, 1999), divides the whole territory into three regions:
- southeastern humid soil region (41.6 % in area);
- central dry and humid soil region (22.7 %);
- northwestern dry soil region (35.7%).

Due to the paucity of information on the actual extent of the monsoonal influence, this book considers the humid areas of China, which form about 50% of the country, to be the monsoonal part of China. Delineation for South and Southeast Asia follows the ordinary geographical concept, that is, the Indian subcontinent, Myanmar and the countries of the Indochinese Peninsula as well as Malaysia, Indonesia and the Philippines in insular Southeast Asia.

2.2.2 Specific features of Monsoon Asia

Using statistics from the United Nations Food and Agriculture Organisation (FAO, 2001) population, land area, and the cultivated area of Monsoon Asia were analyzed and compared with those of the world and Asia as a whole, as shown in Table 2.2. In this calculation, 50% of the land area and 90% of both the population and the cultivated area of China were assumed to be in the monsoon region.

It is clear from the table that Monsoon Asia is a region with an extremely high population density. It has about 3.2 billion inhabitants or some 53% of the world's population, yet it has only 10.5% of the world's total land area. Thus, the population density of Monsoon Asia is as high as 2.25 persons ha^{-1}, or exactly five times that of the world average of 0.45 persons ha^{-1}.

As a corollary, the ratio of the cultivated area to the total land area is 31% for Monsoon Asia, as compared to 11.6% for the whole world. In this

Table 2.2 *Population, land area, and cultivated area of the world, Asia, Monsoon Asia, and tropical Asia (1999)*

	Population		Land area (A)		Cultivated area (B)		B/A
	×10³	(%)	×10³ha	(%)	×10³ha	(%)	(%)
World	5,978,396	(100.0)	13,048,407	(100.0)	1,511,766	(100.0)	11.6
Asia	3,634,278	(60.8)	3,087,109	(23.7)	554,308	(36.7)	18.0
Monsoon Asia	3,176,199	(53.1)	1,374,883	(10.5)	426,822	(28.2)	31.0
Tropical Asia	1,832,816	(30.7)	848,946	(6.45)	295,998	(19.6)	34.9

Note: Tropical Asia corresponds to the difference between the figure of Monsoon Asia and that of the sum of monsoon regions of China, Korean Peninsula and Japan. (Source: FAOSTAT, 2001)

respect, tropical Asia, with lesser climatic and topographic constraints for land use, is even more prominent, with 35% of the total land already under cultivation. With 30% or more as a ratio of cultivated to total land, the statistics would indicate that practically no more land is available for the further expansion of cultivation. Intensification of agriculture is therefore the only choice for increasing food production. However, in some of the countries, the *per capita* area of land under cultivation is as low as 0.17 ha, so the potential to increase production from existing land may also be very limited.

Before dealing with contributing factors that have made Monsoon Asia the rice basket of the world, this section has delineated the geographical extent of Monsoon Asia and identified its characteristic features. The three most outstanding features of Monsoon Asia are:

- a very high population density;
- intensive use of the land for food production;
- the dominance of paddy rice cultivation in its agricultural systems.

2.3 Climate

The very fact that about 90% of the world's rice lands exist in Monsoon Asia suggests the importance of climatic conditions in determining the extent of rice cultivation. As rice is known to tolerate a temperature as low as 13–14° C at the seedling stage, the greater part of the land south of 45° N latitude could support one crop of rice a year. Thus, temperature is not a strong constraint for rice cultivation even for the areas outside of Monsoon Asia.

What is more restrictive for rice cultivation, not only for its total amount, but also for its temporal distribution, is the availability of water. Although lowland rice can be grown on uplands, its performance is better with submergence—at least during the tillering and the heading stages.

2.3.1 General considerations on the water regime of paddy lands
A greater part of Monsoon Asia is governed by a humid to subhumid climate. Much of East Asia and equatorial tropical Asia experience limited severe drought conditions throughout the year. In tropical Southeast and South Asia, the regions at a distant from the equator have a long dry season during the winter months. However, even these regions have quite a high rainfall brought by the monsoon during the summer months.

The water requirement for lowland rice is determined by evapo-transpiration and leaching during the growing season, as well as the water required for land preparation and nurseries. This requirement is estimated to be about 1000 mm for paddy soils with almost no downward percolation, and 1500 mm or more for those with moderate percolation (as frequently found in Japan). For most purposes, 1000 mm per crop would be the minimum water requirement. If this quantity of water has to be supplied solely by rain, the rainfall over 4–5 months, depending on the varieties, must exceed 1000 mm. If the percentage of rain falling during the rainy season is assumed to be 80%, the annual rainfall of the area must exceed about 1250 mm.

The assumption that 80% of the total rainfall is concentrated in the rainy season may be justifiable for most regions with a tropical monsoon climate. If 1250 mm is set as the annual rainfall threshold, most rice-growing areas in tropical Asia have more rain than the threshold, while non-rice-growing areas have less. Thus, the threshold value of 1250 mm appears reasonable.

If there are other sources of water for irrigation, areas with much lower rainfall may be cultivated for rice. By contrast, if the concentration of rainfall is not sufficiently high, those areas with much higher annual rainfall cannot grow rice by relying solely on rainwater. The latter case may not be a normal occurrence because with reasonably abundant rainfall it is relatively easy to secure water for irrigation.

2.3.2 Water balance study on paddy soils of tropical Asia

Kyuma (1973) studied the water balance of paddy soils in tropical Asia using his own scheme of climatic regional division (Kyuma, 1972) as the frame of reference (as shown in Figure 2.1), and a modified method of Thornthwaite's water balance calculation (Thornthwaite, 1948).

For a purely rain-fed condition, Kyuma made the following assumptions for his calculations:

- Soil can retain 100 mm of water in the effective rooting depth of the soil profile, as assumed by Thornthwaite;
- If the bund height is 20 cm, a maximum of 200 mm of water can be retained and ponded on the field while any excess is lost as run-off;
- Percolation loss from the ponded water is a maximum of 100 mm per month, which corresponds to a hydraulic conductivity of about 4×10^{-6} cm/sec.

The surplus or deficit of water was calculated for each month from the difference between the actual precipitation and Thornthwaite's potential evapotranspiration, using the above assumptions. An example of the computation is given in Table 2.3.

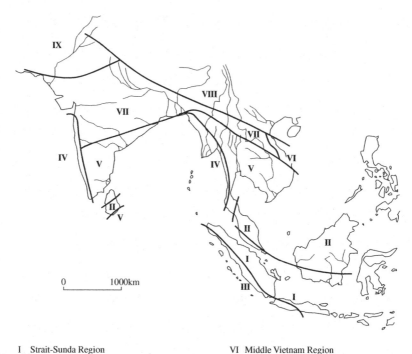

I   Strait-Sunda Region   VI   Middle Vietnam Region
II   Malay-Northern Borneo Region   VII Central India-Northern Indochina Region
III Oceanic Sumatra-West Java Region   VIII Tongking-Assam Region
IV   Southwest-Facing Coastal Region   IX   Lower Indus Region
V   Southern Indochina-Southern India Region

Figure 2.1
Map of tropical Asia showing climatic regions (Source: Kyuma, 1972)

The results are summarized in Table 2.4 and Table 2.5, and some examples of water regime patterns are illustrated in Figure 2.2 for stations selected from each of the climatic regions established by Kyuma. In Table 2.4 the stations are grouped according to the available water regime, based on the number of consecutive months over which soil moisture is available. It is obvious from the table that the climatic Regions V, VII, and IX have less favorable water regimes than the other regions. As seen in Figure 2.1, this is not surprising for Region IX, which is the Lower Indus Region and is classified either as arid or semiarid by Thornthwaite's method. However, it is rather surprising to find that a fairly large area of Region V, the Southern Indochina – Southern India Region, and Region VII, the Central India –

Table 2.3 Computation of the soil water regime for Yangon (in mm)

Month	1	2	3	4	5	6	7	8	9	10	11	12	Annual
Pot. evapotrans.	92	100	145	167	176	158	162	162	147	152	138	103	1692
Rainfall	8	5	6	17	260	524	492	492	398	208	34	3	2529
Soil Moisture	0	0	0	0	84	100	100	100	100	100	100	0	–
Ponded water	0	0	0	0	0	200	200	200	200	156	0	0	–
Percolation	0	0	0	0	0	100	100	100	100	100	52	0	552
Runoff	0	0	0	0	0	50	230	230	151	0	0	0	753
Deficiency	84	95	139	150	0	0	0	0	0	0	0	0	468

(Source: Kyuma, 1973)

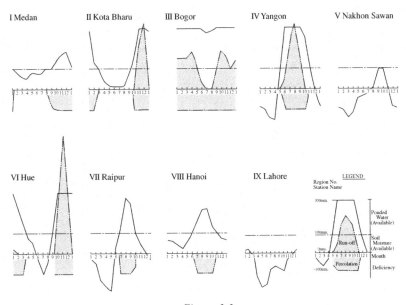

Figure 2.2
Patterns of soil water regime for climatic regions I–IX (Source: Kyuma, 1973)

Northern Indochina Region, have water regimes so severe that they do not seem to be able to support even a single short-term upland crop.

Table 2.5 shows the number of stations for each climatic region in each category of ponded water regimes. For this table, a water regime that has two or less consecutive months of ponded water is thought to be prohibitive of rice cultivation. A regime having three consecutive months of ponded water is regarded as marginal. Four to six months of ponded water may be

Table 2.4   Grouping of sample stations according to climatic region and available water regime

Climatic region	Available water regime lasting consecutively for				Total
	≤3 months	4–6 months	7–9 months	≥10 months	
I	1	2	5	9	17
II	0	1	1	10	12
III	0	0	0	8	8
IV	0	3	11	1	15
V	10	11	11	0	32
VI	0	0	6	0	6
VII	4	5	11	1	21
VIII	0	0	2	6	8
IX	6	0	0	0	6
Total	21	22	47	35	125

(Source: Kyuma, 1973)

Table 2.5   Grouping of sample stations according to climatic region and ponded water regime

Climatic region	Ponded water regime lasting consecutively for				Total
	≤2 months	3 months	4–6 months	≥7 months	
I	6	3	7	1	17
II	0	2	3	7	12
III	0	0	0	8	8
IV	0	0	12	3	15
V	25	4	3	0	32
VI	0	0	5	1	6
VII	12	4	5	0	21
VIII	1	1	4	2	8
IX	6	0	0	0	6
Total	50	14	39	22	125

(Source: Kyuma, 1973)

sufficient for one crop of rice. Ponded water regimes lasting over six months would have some problems in drainage.

Using the results given in Tables 2.4 and 2.5, the following remarks may be made about each climatic region.

Regions I, II, III, IV, VI, and VIII are generally suitable for rice cultivation, even under rain-fed conditions.

Region I, the Strait–Sunda Region, is peculiar in that it has periods where surplus water is available for a considerable length of time, but ponded water only lasts for a short period, or not at all. This type of water regime seems to be common in the southern half of the west coast of West Malaysia and is better suited to multiple cropping of upland crops. Furthermore, within region I a sub-region can be demarcated, separating the northern coast and the eastern half of the island of Java along with the area further east, where a distinct dry season lasts for at least four months. However, even here one crop of rice, depending solely on rainwater, is possible.

Region II (Malay –Northern Borneo) seems to be suitable for one crop of rice during the rainy season and one upland crop for the rest of the year. In some areas, cultivation of the second crop may be somewhat risky if no supplementary irrigation is provided.

Region III (Oceanic Sumatra – West Java) has plenty of rain throughout the year and double cropping of rice should present no difficulties, even without a large-scale irrigation scheme. In this region, upland crop cultivation may face drainage rather than irrigation problems.

The greater part of Region IV (Southwest – Facing Coast) experiences a distinct dry period lasting for more than four months. One crop of rice can be safely cultivated, but a second crop is not cultivable unless some form of irrigation is provided.

Region VI (Middle Vietnam) is similar to Region IV in its soil water regime, but the dry period is not as severe.

Region VIII (Tongking–Assam) has a favorable water regime for one crop of rice and may be followed by the planting of an upland crop. Here, low temperatures during the winter months pose some difficulties for the cultivation of certain crops.

In contrast to the above regions Regions V, VII and IX are generally not suited to rice cultivation, due to their unfavorable soil water regimes. Twenty-nine out of thirty-two stations in Region V, sixteen out of twenty-one in Region VII, and all six stations in Region IX, are either prohibitive or marginal for rice cultivation under rain-fed conditions. Important rice areas occurring in these regions are either completely irrigated, as in Region IX, or located in water conservation areas, mostly at the mouths of big rivers where the land is naturally inundated due to landform conditions for three to four months during the cropping season. However, the greater part of the rice lands in Cambodia, Thailand and India, countries that belong to Regions V and VII, are neither completely irrigated nor located within water conservation areas, and therefore have a very unstable water supply. A considerable year-to-year fluctuation in the area planted to rice and of large areas with droughts or floods, may be cited as evidence of the instability caused by unfavorable soil water regimes.

The preceding computations and discussion are based on a long-term average of meteorological data. It is well known that the monsoon tropical climate is quite variable, and the timing and the amount of rainfall can fluctuate greatly from one year to another. Thus, the soil water regime as illustrated in Figure 2.2 should be regarded as one of the more frequent patterns occurring at the site.

2.4 Landforms

As discussed above, climatic conditions alone do not explain the present distribution of rice cultivation. The actual water conditions in rice growing lands are also governed, to a greater extent, by landforms, as evidenced by the water conservation (or naturally inundated) areas in climatic Regions V and VII. Actually, the vast extent of lowlands is an even more prominent feature of Monsoon Asia than its climate.

2.4.1 Occurrence of extensive lowlands in Monsoon Asia

The map of Monsoon Asia shows many vast rivers with large alluvial plains and deltas, including the Yangtze, Mekong, Chao Phraya, Irrawaddy, Ganges, and the Brahmaputra. The existence of these huge rivers and of the extensive alluvial lowlands along their middle and lower reaches is a notable characteristic of the landforms of Monsoon Asia.

Table 2.6 provides data on the area of alluvial soil and the total land area of the world, Asia and tropical Asia. For the world as a whole, alluvial soils occupy less than 1/20 of the total land area, while in tropical Asia they occupy 1/6 of the area. Potentially arable land in alluvial soil areas is

Table 2.6 Importance of alluvial soil area in Asia and tropical Asia (in 10^6 ha)

| | Land area | | Alluvial soil area | |
	Total	Potentially arable	Total	Potentially arable
World	13000	3152	588	316
Asia*	2704	620	–	192
Tropics	4893	1652	365	172
Tropical Asia	987	344	168	114

* Excluding former USSR
(Source: White House, 1967)

only 1/10 of that in the whole world, while it is 1/3 of that in Asia and tropical Asia.

In other words, more than 1/3 of the world's potentially arable land in alluvial soil areas is located in tropical Asia, which itself occupies only 1/13 of the world's total land area. Also, almost 2/3 of all potentially arable lowlands in the tropics are concentrated in tropical Asia, which has only 1/5 of the total land area of the tropics. Africa and America have much smaller shares of lowlands in their tropical regions.

From this information; one can conclude that the use of alluvial lowlands is an absolute necessity in Asia and tropical Asia, where 1/3 of the potentially arable land is in such lowlands. The adoption of rice as a crop may be regarded as an adaptation to the landform conditions in Monsoon Asia. In a sense, rice cultivation in Monsoon Asia is not necessarily a result of human preference, but a result of nature's enforcement.

Thus, the unique combination of the monsoon climate and the exceptionally large lowland area that it covers is the driving force that makes Monsoon Asia the rice basket of the world.

2.4.2 Geological considerations on the extent of the lowlands

The climate of Monsoon Asia is not unique to this region. Other monsoon systems occur, for example, in West Africa. However, the abundance of lowlands is more specific to Monsoon Asia. Why, then, is Monsoon Asia endowed with such an exceptionally large area of lowlands? To answer this question the geology and geohistory of Monsoon Asia need to be examined.

Many of the continents today originated from an ancient Paleo–Mesozoic super-continent, 'Gondwanaland', which started to fragment during the mid-Mesozoic era, about 130 million years ago. Africa was at the core of Gondwanaland. South America, Australia, and Antarctica gradually drifted away. Asia also had some share, such as the Arabian Peninsula and the Deccan Plateau of India. All of these former components of Gondwanaland are dominated by the shield structure, which is a very stable landmass and has not been subjected to active crustal movements. Therefore, the land surface has undergone thorough weathering for millions of years and has slowly been denuded to form a planation surface or peneplain. Under a humid climate, deeply weathered, acidic and very infertile red soils predominate in the gently undulating landscape of such a peneplain.

In contrast to the geologically stable continents that originated from Gondwanaland, the main body of Monsoon Asia is characterized by its

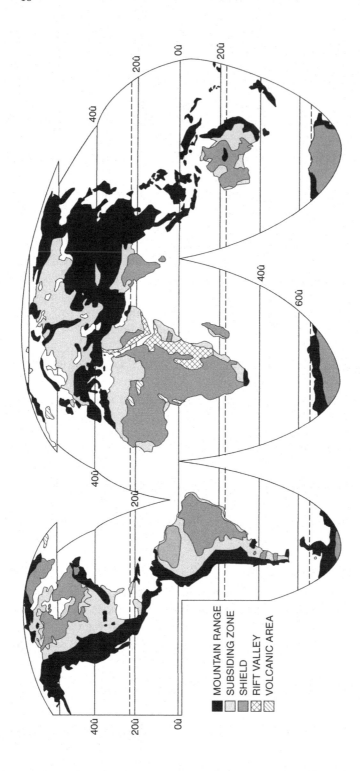

Figure 2.3
World structural regions (Source: Murphy, 1968)

geological instability, accompanying intensive orogeny and volcanism. The uplifting of the Himalayas and associated mountain ranges occurred in its continental component and volcanic activities continued along the Circum–Pacific volcanic zone in its insular component. Figure 2.3, taken from Murphy (1968), shows the difference in geological structure between Asia and the other continents. Figure 2.4 shows a high frequency of earthquakes with shallow foci in Monsoon Asia, further emphasizing the geological instability of the region.

Uplifting mountains and newly erupted volcanic ejecta are subjected to the strong erosive forces of heavy monsoon rains. A vast amount of water during the rainy season must be drained by many, large rivers. These rivers transport a huge amount of eroded sediment and deposit it on the alluvial plains in the middle and lower reaches and into deltas at the river mouths, resulting in an exceptionally high share of lowlands in Monsoon Asia. In addition, these alluvial plains and deltas are also endowed with very fertile soils derived from relatively unweathered materials.

As discussed already, the abundance of lowlands favorable to paddy rice cultivation in Monsoon Asia is controlled geologically. Fertile soils and

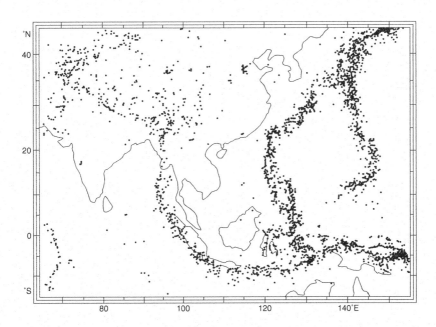

Figure 2.4
Earthquakes with a focus within 100 km (1961–1967)
(Source: Barazangi and Dorman, 1969)

rich water resources are the advantages complementing the physiographic condition.

2.4.3 Landforms of rice lands and their relation to rice cultivation

Landforms are crucial to the successful cultivation of rice, and the situation in Thailand is discussed here as an example. Takaya (1971) made a landform study of rice lands in the Chao Phraya Basin of Thailand. He defined six physiographic regions: intermontane basins, the constricted river channel area, the old delta, the delta flat, the deltaic high, and fan-terrace complexes, as seen in Figures 2.5 and 2.6. The characteristics of these six physiographic regions are summarised in Table 2.7. Fukui (1971) superimposed his rice cultural regions on Takaya's physiographic regional division and discussed the features of rice cultivation in each region. What follows next is a brief description of each of the physiographic regions set up in Thailand in relation to rice cultivation, based on the studies by Takaya and Fukui.

The intermontane basins are situated in Northern Thailand. They are characterized by high elevation, steep general slopes, strong local relief, and a small area. The complexity of landforms or the occurrence of different physiographic units, such as recent alluvial plains, recent fans, fan-colluvial complexes and low and high terraces within an intermontane basin is in striking contrast to the other more homogeneous physiographic regions. In this physiographic region, most paddy fields are found either on recent alluvial plains, low terraces or recent fans. The high catchment to paddy area ratio is favorable for irrigated paddy cultivation. Further, the small size of the rivers and the moderate general slope make water management feasible at a communal level. Consequently, rice cultivation in the intermontane basins is thought to have started in early times.

The constricted river channel area comprises elongated depressions along the Nan and Yom rivers with an average width of about 20 km. This area is characterized by annual prolonged deep flooding, which is caused by the bottle-necked gorge on the Chao Phraya River near the confluence of the Nan and the Ping rivers. The prolonged deep flood during the rainy season presents a severe limitation to land use. Floating rice is the only cultivable variety in most areas of this region. Drainage rather than irrigation is the constraint.

The old delta region occupies the upper part of the present-day Bangkok Plain. The Chao Phraya's distributaries start to branch off near Chai Nat and the deltaic nature of the plain is obvious. Geomorphologically, this is an upper Pleistocene delta partially covered by recent levee deposits. The

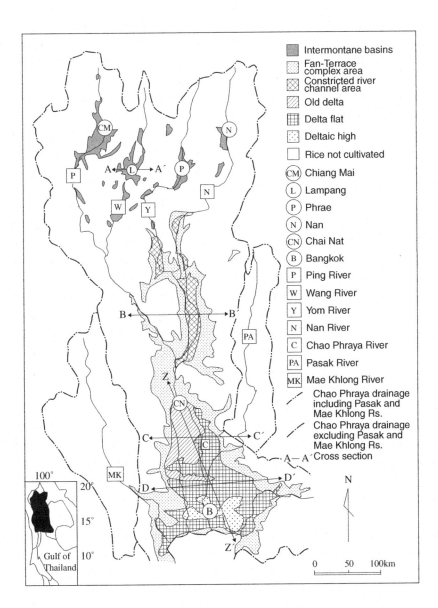

Figure 2.5
Physiographic classification of paddy land of the Chao Phraya River basin
(Cross-section and profile are shown in Figure 2.6.) (Source Takaya, 1971)

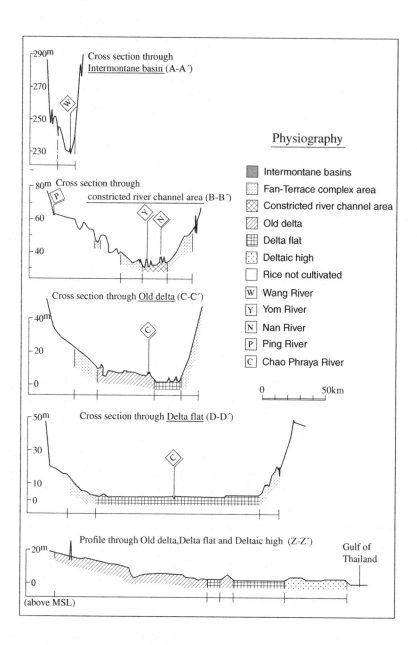

Figure 2.6
*Cross-section and profile of the Chao Phraya River basin (The locations are shown
in Figure 2.5.) (Source: Takaya, 1971)*

Table 2.7 *Characteristics of the six physiographic regions*

Physiographic Region	Elevation m	General slope m km⁻¹	Local relief m	Acreage (×1000ha)		Soil texture	Geomorphic setting	Catchment/ paddy area
				Gross	Net			
Intermontane basin	150–350	>1.7	<10	250	200	clay to gravelly sand	Complex of stream alluvium, fan and terraces	10–40
Constricted river channel area	23–60	0.2	±5	320	200	clay with a little sand	Recent alluvial plain	32.8
Old delta	5–20	0.15	<8	490	400	clay with a little sand	Upper Pleistocene delta	23.0
Delta flat	0–2	0.01	negligible	1110	820	clay	Interdistributary low of recent delta	15.6
Deltaic high	2–4	0.01	negligible	210	160	clay with a little sand	Relatively elevated parts in recent delta	
Fan-terrace complex area	5–100	1.0–2.5	<10	1800	1350	sand with a little clay and gravel	Recent and Pleistocene fans and terraces	5.1

(Source: Takaya, 1971)

general slope is very gentle, but local relief is moderate, reflecting the normal dissection and the superimposition of recent deposits. Because of the local relief, some parts of the region are flooded too deeply and some parts are not submerged naturally. Today this region is under a government-administered large-scale irrigation scheme, and the greater part has become cultivable for rice, even for double cropping if sufficient water is available. However, water conditions in the depressions, have not been improved and only traditional varieties of floating rice can be cultivated

The delta flat region is a recent delta formed by the Chao Phraya and has a very flat surface with negligible local relief. At one time, this region resembled a large lake during the rainy season, but became a dry plain during the dry season. Since the mid-nineteenth century, many canals have been dug to make this area inhabitable, and as a result, it has become the rice bowl of Thailand. However, because of the widespread occurrence of brackish sediments, the greater part of the area (about 800,000 ha) has acid sulfate soils (see Chapter 11).

The fan-terrace complex region occurs along the margins of the Central Plain, where the plain proper merges into the mountainous region. Geomorphologically the area is composed of recent and upper Pleistocene fans and terraces, some of the material as well as topographic features of the latter are illustrated in Figure 2.7 along with those of even older sediments. What is most characteristic of the region as a whole is the low catchment to paddy area ratio, which is estimated to be around five. Therefore, the area is basically water-deficient under the prevailing climatic conditions of Thailand. Where the area of paddy fields is small, the terrain is similar to that in the intermontane basins with respect to the ease of water management. Thus, some of the fans have a long history of rice cultivation and very old water distribution system. However, as the paddy area was expanded, water deficiency became acute, and the greater parts of rice lands in this region today are rain-fed. The conditions described previously for the soil water regime of Region V apply to this fan-terrace complex region. According to Fukui's estimate (1971), the area of this landform region amounts to 1.3 million hectares, or about 40% of the paddy lands in the Chao Phraya River Basin.

Most of the physiographic regions delineated in the Chao Phraya River Basin also appear in other major river basins. Even the constricted river channel area, which seems to be determined by a particular geologic structure in Central Thailand, has its counterparts in other regions, such as Northeast Thailand along the Mun, and Assam along the Brahmaputra. Only the deltaic high region is not encountered as an independent unit in

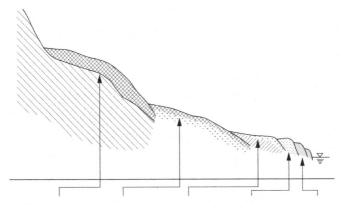

FORMATION	PENEPLAIN	TERRACE III	TERRACE II	TERRACE I	FLOOD PLAIN
SEDIMENT	No young alluvium	Dominantly sandy	Dominantly clayey	Loamy	Clayey to sandy
WEATHERING FEATURE	Thick hard laterite	Heavily weathered alluvium capped with laterite	Weatherd alluvium with pisolitic concretions	Alluvium with soft iron concentrations	Fresh alluviun with iron mottles
TOPOGRAPHY	Undulating to rolling	Slightly undulating to undulating	Flat to slightly undulating	Flat	Flat and slightly undulating
AGE	Lower Pleistocene	Middle Pleistocene	Upper Pleistocene	Holocene or Uppermost Pleistocene	Holocene

Figure 2.7
Schematic cross-section of Quaternary deposits in Thailand
(Source: Takaya, 1968, partly modified)

other deltas. Thus the delta flat and the deltaic high, should be treated as one—the young delta region—in a general discussion.

Although the fan-terrace complex region occurs extensively throughout tropical Asia, its potential for rice production varies from one climatic region to another. In Regions V and VII water deficiency is acute, as in Thailand, so rice cultivation is inevitably unstable. However, in regions III and IV, for example, the fan-terrace complex region receives abundant rains during the growing season, so even rain-fed rice cultivation can be quite stable.

In order to make the discussion comprehensive, it may be necessary to add a few more physiographic regions. Takaya (1972) and Takaya and Tomosugi (1972) proposed two additional regional categories, that is, coastal region and plateau region.

The coastal region is a thin strip of young delta along the coast and has three sections: beach ridges, swales between beach ridges, and silted lagoons. Due to its low-lying topographic position, this region is inundated for a considerable period of the year, but not too deeply because water is freely drained into the nearby sea. Rice can be grown without sophisticated water control measures. Possible problems for rice cultivation in the coastal region are the brackish nature of the deposits that may give rise to acid sulfate soils when they are reclaimed, and direct salinization by sea water.

The plateau regions occur in Northeast Thailand, Upper Myanmar, and India. They are dissected Plio-Pleistocene surfaces or uplifted peneplains, and usually have undulating to rolling topography. Rice is grown in the depressions. As the catchment to paddy area ratio is even lower than that of the fan-terrace complex region, climatic control on the water conditions of their rice lands is more direct. The fact that these plateau regions are all in climatic Regions V and VII explains the very unstable rice cultivation in these regions.

2.5 Soil materials of rice lands

2.5.1 General characteristics
The soils of rice lands occur mostly on alluvial landforms of the Holocene and upper Pleistocene periods, such as floodplains, deltas, and lower terraces. Therefore, they are not well developed in morphologically, and the majority is classified as Entisols or Inceptisols in Soil Taxonomy (Soil Survey Staff, 1999). Some soils occurring on terraces may have clay translocation within the profile and can be classified as Alfisols or Ultisols, depending on the base status. In addition to their occurrence on low-lying lands, high rainfall with a heavy concentration in a relatively short period of time causes natural inundation or floods. Thus, signs of wetness, such as gleization, mottles and concretions, or redoximorphic features using Soil Taxonomy terminology, are almost always seen even under the natural condition prior to the cropping of rice. Sometimes even unripe gleyed sediments are seen underlying a shallow solum.

It is pertinent here to point out that these alluvial sediments were originally derived from the richest part (the topsoil) of mountain and upland soils that underwent intense chemical weathering under the humid tropical climate, so are normally very poor in nutrient reserves. Therefore, alluvial soils are by and large better off in terms of soil chemical and mineralogical properties compared with upland and mountain soils. In addition, it is possible that some rejuvenation might have occurred since the sediments were deposited, particularly in the deltaic environment. Even so, the soil chemical characteristics are generally governed by the nature of original sediments, because the soil has had insufficient time to fully mature.

Paddy soils in the lowlands are therefore usually acidic in reaction and are often characterized by the relative dominance of 0.7 nm or kaolin minerals in the clay fraction. The texture varies widely depending on the characteristics of the catchment and the sedimentation process. However, as they are often found in wide alluvial and deltaic plains, the soil texture tends to be medium to fine.

2.5.2 Soil texture and particle size distribution

Soil material and fertility characteristics of tropical Asian paddy soils are shown in the following tables and figures that use data from Kawaguchi and Kyuma (1977) and Kyuma (1985). Table 2.8 compares the particle size distribution of tropical Asian paddy soils to that of Japanese paddy soils. The distribution of individual samples from the respective countries is also shown in the triangular diagram for the Japanese system of soil texture classification, as shown in Figure 2.8.

The abundance of heavy-textured soils is noticeable in Indonesia, the Philippines, East and West Malaysia, of which the first two are in the volcanic regions of insular Southeast Asia. The basic nature of volcanic ejecta allows

Table 2.8 Country means and standard deviations of sand, silt and clay contents

Country	No. of samples	Sand, %		Silt, %		Clay, %	
		Mean	Std. Dev.	Mean	Std. Dev.	Mean	Std. Dev.
Trop. Asia	529	23.3	–	30.5	–	41.2	–
Bangladesh	53	25.6	20.0	42.8	13.1	31.6	17.2
Myanmar	50	14.7	14.1	42.7	11.1	42.6	18.5
Cambodia	16	33.9	25.5	31.4	12.7	34.7	24.3
India	73	36.4	22.8	24.3	11.2	39.3	19.6
Indonesia	44	22.5	18.3	26.3	10.8	51.2	24.2
E. Malaysia	36	14.4	18.1	39.2	12.9	46.4	16.6
W. Malaysia	41	22.5	22.0	31.3	10.6	46.1	16.1
Philippines	54	27.1	20.8	30.8	9.2	42.1	18.8
Sri Lanka	33	68.3	20.2	7.6	6.0	24.1	15.8
Thailand	80	38.2	29.3	25.2	11.1	36.7	24.6
Vietnam	49	9.4	14.4	34.6	8.3	56.0	12.0
Japan	155	49.2	18.2	29.6	10.6	21.2	10.1

Notes: Myanmar—mainly from the Irrawaddy Delta; India—from the states along the Ganges and the east coast of Deccan; Indonesia—Java; E. Malaysia—Sarawak; Vietnam—Mekong Delta; for others more or less from throughout the country. (Source: Kyuma, 1985, partly modified)

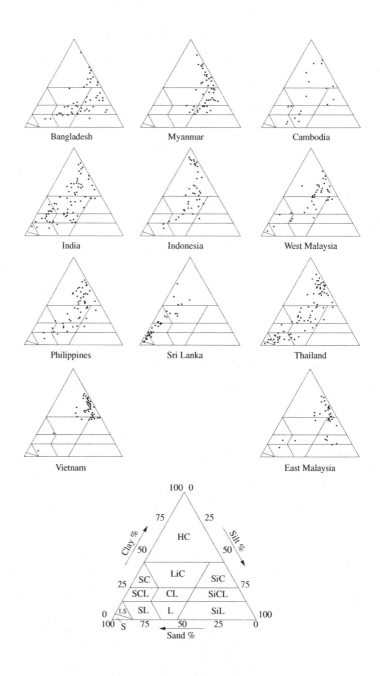

Figure 2.8
Distribution of sample soils in the triangular diagram for textural classification
(Source: Kyuma, 1985)

ready weathering to clay without leaving sand-sized particles. Many of the rice areas in East and West Malaysia are reclaimed from brackish and freshwater swamps, which are rich not only in organic matter but also in clay.

Among the Thai soils, sandy light-textured soils from the Northeast or Khorat Plateau region make one cluster, as seen in the lower left corner of the triangle. Another large cluster of samples, seen in the upper right corner, is mainly composed of soils from the Bangkok Plain.

The heavy-textured soils from India are mostly Vertisols or vertic alluvial soils in the deltas of such major rivers as the Ganges, Godavari–Krishna, and the Cauvery. Many of the sandy clay soils and the sandy clay loam soils are on strongly weathered old alluvial terraces or laterite-capped plateaus.

Soils of Bangladesh and Myanmar are characterized by their high silt content, though the latter are also high in clay. The samples of both countries are dominated by the alluvial sediments of such major rivers as the Ganges–Brahmaputra and the Irrawaddy. In contrast, Sri Lankan samples are sandy and very low in silt, which may be because the greater part of paddy soils in Sri Lanka are developed on residual or local alluvial material, that is, poorly sorted materials derived from well-weathered gneissic rocks.

Compared with Japanese paddy soils, the mean clay content of tropical Asian paddy soils is much higher, reflecting difference in geomorphic processes. By and large, alluvial plains and deltas in tropical Asia are large and flat, so fine materials are deposited in alluvial lowlands. However in Japan, rugged mountains dominate, and alluvial lowlands are situated either in narrow coastal strips or in small intermontane basins. Consequently, even Japan's alluvial lowlands are made up of a large proportion (about 60%) of alluvial fans with coarser grain-size compositions (Mitsuchi, 1981).

2.5.3 Soil clay mineralogical composition

Clay mineralogical composition was semi-quantitatively assessed by measuring the areas of 0.7nm (kaolinite, halloysite group), 1.0nm (illite, or clay mica), and 1.4nm (smectite and vermiculite group) peaks on x-ray diffractograms and by calculating the relative content (%) of each in crystalline clay minerals. The results are given in Table 2.9 and Figure 2.9, the latter showing the distribution of individual samples in a triangular diagram.

The high relative content of 1.4 nm minerals, mainly composed of smectites, among crystalline clay minerals, was rather unexpected. They were particularly high in the volcanogenous soils of Indonesia and the

Philippines, and also in the deltaic sediments. This was reflected by the high cation exchange capacity (CEC) values of these soils (see Table 2.10). It is often said that low activity, variable charge clays dominate in tropical soils, but this does not apply to paddy or lowland soils. This is a very important point when considering the fertility management of paddy soils.

There appear to be roughly three types of clay mineral composition:

- 0.7nm—dominant type: Cambodia, West Malaysia, Sri Lanka and Thailand
- 1.4nm—dominant type: Indonesia and the Philippines
- 0.7–1.0–1.4nm—even type: Bangladesh, Myanmar, India, Vietnam and East Malaysia

Of the three types, the 0.7 nm—dominant type may be regarded as the most highly weathered and the 1.4 nm—dominant type as the least weathered, with the 0.7–1.0–1.4nm—even type being intermediate.

Samples containing more than 75% of 1.4nm minerals are either Vertisols or vertic alluvial soils derived from calcareous sediments or from basic volcanic ejecta. The 1.4–0.7 and 0.7–1.4 combinations occur most frequently among the soils developed on recent alluvial and deltaic sediments. Only Bangladeshi soils have an appreciable number of 1.4–1.0 combinations, and the soils with this clay composition are exclusively of Gangetic alluvia origin.

Soils containing more than 75% 0.7nm minerals are most frequently encountered in the wet zone of Sri Lanka, the east coast of West Malaysia

Table 2.9 Country means and standard deviations of clay mineral composition

Country	No. of samples	0.7nm, %		1.0nm, %		1.4nm, %	
		Mean	Std. Dev.	Mean	Std. Dev.	Mean	Std. Dev.
Trop. Asia*	529	44.0	–	16.8	–	39.2	–
Bangladesh	53	34.3	10.7	29.2	14.4	36.5	13.6
Myanmar	50	26.9	7.4	17.3	6.5	55.8	10.7
Cambodia	16	60.6	19.2	4.1	8.5	35.3	17.8
India	73	34.7	16.4	23.7	8.2	41.5	20.6
Indonesia	44	46.6	29.7	0.7	15.3	52.7	30.5
E. Malaysia	36	35.2	14.2	33.5	17.6	31.3	12.7
W. Malaysia	41	59.4	22.2	9.0	2.6	31.6	23.4
Philippines	54	31.2	19.4	2.6	4.5	66.2	24.3
Sri Lanka	33	64.2	25.6	9.1	11.5	26.7	23.6
Thailand	80	59.2	16.5	12.9	12.0	27.9	15.5
Vietnam	49	47.2	9.4	33.1	7.7	19.8	5.2

* Refer to the notes for Table 2.8.
(Source: Kyuma, 1985, partly modified)

and the southern peninsular region of Thailand. Many sample soils from the Khorat Plateau region of Thailand and Cambodia also have this clay composition. Besides these regions, soils containing a high amount of kaolin minerals are commonly found on highly weathered middle and high terrace sediments, which often contain pisolitic concretions or lateritic nodules.

2.5.4 Soil fertility characteristics

Various fertility characteristics of tropical paddy soils are listed as country means and compared with those of Japanese paddy soils in Table 2.10. Only some of the points worthy of note are discussed.

Although the mean pH of tropical Asian paddy soils is higher than Japanese paddy soils, some of the country means are definitely low, such as those of East and West Malaysia, Myanmar, and Vietnam. Malaysian paddy soils often contain large amounts of peaty organic matter, which may have lowered the mean pH. The samples from Myanmar and Vietnam were derived from deltaic sediments of the Irrawaddy and the Mekong, respectively. As is discussed in Chapter 7, the pH of brackish deltaic sediments is considered to converge to between 4.5–5, and this would apply to these two country means. The soils of Bangladesh are also derived from deltaic sediments, but the calcareous nature of the Gangetic alluvia, which constitute an important part of parent materials of Bangladeshi paddy soils, might indicate an exception.

Organic matter is clearly lower in tropical Asian paddy soils than Japanese paddy soils, with the exception of Malaysian soils. However, even the latter show relatively low nitrogen availability, probably because of the peaty nature of their organic matter.

The most remarkable difference between tropical paddy soils and Japanese paddy soils is seen in their phosphorus status, both total and available. This may have resulted from differences in fertilizer application. Among tropical Asian paddy soils, volcanogenous soils in Indonesia and the Philippines appear to be higher in total phosphorus, but their available phosphorus is not necessarily high.

As mentioned earlier, the CEC of tropical Asian paddy soils is generally high, reflecting both a high clay content and a relatively high proportion of 1.4 nm minerals in their clay mineral composition. Paddy soils in the tropics cannot be grouped with many of the tropical upland soils under the category of low activity clay soils.

Exchangeable cation composition and available silica content are also related to clay content and clay composition. High base and high available

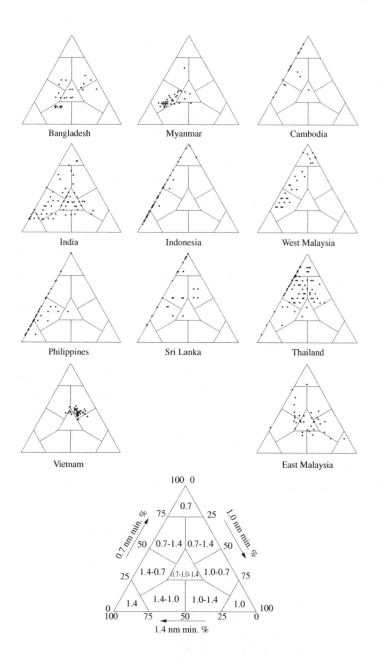

Figure 2.9
Distribution of sample soils in the triangular diagram for clay mineral composition
(Source: Kyuma, 1985)

Table 2.10 *Country means of fertility characteristics of tropical Asian paddy soils*

Country*	No. of samples	pH	TC %	TN %	C/N	NH$_4$-N mg kg⁻¹	NH$_4$-N/TN %	TP	Bray-P P$_2$O$_5$ mg kg⁻¹	HCl-P	CEC cmol kg⁻¹	Exch.Cations, cmol kg⁻¹ Ca	Mg	Na	K	Avail. Si O$_2$ mg kg⁻¹
Trop Asia	529	5.6	2.07	0.17	11.5	96	6.2	821	30	172	19.1	9.3	5.6	1.5	0.4	237
Bangladesh	53	6.1	1.18	0.13	8.7	61	4.5	842	38	210	12.9	7.8	2.7	0.9	0.3	129
Myanmar	50	4.8	1.30	0.12	11.3	35	2.9	582	10	78	20.5	7.2	8.4	1.5	0.4	133
Cambodia	16	5.2	1.09	0.10	10.7	40	5.1	443	6	19	14.6	5.4	3.2	0.3	0.2	113
India	73	7.0	0.85	0.08	11.7	27	3.9	920	81	219	22.0	15.0	6.5	2.4	0.5	347
Indonesia	44	6.6	1.39	0.12	12.6	141	12.4	1337	24	100	26.1	17.8	6.3	1.5	0.4	629
E. Malaysia	36	4.5	9.66	0.64	14.7	302	5.8	1015	11	50	23.0	3.5	3.8	1.2	0.4	103
W. Malaysia	41	4.7	3.36	0.28	11.8	149	6.3	803	37	82	15.9	3.9	5.2	1.5	0.4	104
Philippines	54	6.4	1.66	0.15	11.7	172	10.8	1136	37	134	27.0	14.8	9.3	2.2	0.5	454
Sri Lanka	33	5.9	1.41	0.13	10.5	84	7.0	725	14	90	11.5	5.4	3.5	0.5	0.2	216
Thailand	80	5.2	1.05	0.09	11.3	52	6.6	442	14	47	14.4	7.2	4.3	1.4	0.3	122
Vietnam	49	4.5	2.49	0.20	11.9	77	3.7	764	12	51	18.8	7.1	5.4	1.2	0.4	154
Mediter.	62	6.8	1.82	0.16	10.6	76	4.9	1038	72	–	18.0	15.9	4.6	1.0	0.6	239
Japan	84	5.4	3.33	0.29	11.6	175	6.5	2204	129	465	20.3	9.3	2.8	0.4	0.4	195

* Refer to the notes for Table 2.8; Mediter. is the means for paddy soils in Portugal, Spain and Italy.
(Source: Kawaguchi and Kyuma, 1977, partly modified)

silica status are particularly conspicuous for volcanogenous soils in Indonesia and the Philippines.

Generally speaking, tropical Asian paddy soils, except for some of the extremely weathered soils like those from the Khorat Plateau of Thailand, are not inferior in their fertility characteristics to paddy soils in Japan. Doubtless, differences in management practices for many years may have created some disparities between tropical and Japanese soils, such as those in nitrogen and phosphorus status, but that does not mean that there is any substantial difference in their inherent soil quality. There should be no difficulty in improving the fertility status of soils to a level that would allow much higher rice yields.

References

Barazangi, M. and Dorman, J. 1969. World seismicity maps compiled from ESSA, Coast and Geodetic Survey, Epicenter Data, 1961–1967. *Bull. Seismol. Soc. America*, 59 (1): 369–380. (Cited from Hotta, M. 1989. III. Forest and man. *Transactions of A Colloquium 'Southeast Asia and Africa: Towards Inter-Area Studies'*. Coordinating Group for Overseas Researches).

Food and Agriculture Organization 2001. FAOSTAT. FAO, Rome.

Fukui, H. 1971. Environmental determinants affecting the potential dissemination of high yielding varieties of rice—A case study of the Chao Phraya River basin. *Tonan Ajia Kenkyu (Southeast Asian Studies)*, 9: 348–374.

Gong Zitong 1999. *Systematic Classification of Soils in China*. Science Publisher, Beijing. (In Chinese).

Greenland, D.J. 1997. *The Sustainability of Rice Farming*. CAB Int'l, U.K. & IRRI.

Kawaguchi, K. and Kyuma, K. 1977. *Paddy Soils in Tropical Asia, Their Material Nature and Fertility*. Univ. Press of Hawaii, Honolulu.

Kyuma, K. 1972. Numerical classification of climate of south and southeast Asia. *Tonan Ajia Kenkyu (Southeast Asian Studies)*, 9: 502–521.

Kyuma, K. 1973. Soil water regime of rice lands in south and southeast Asia. *Tonan Ajia Kenkyu (Southeast Asian Studies)*, 11: 3–13.

Kyuma, K. 1985. Fundamental characteristics of wetland soils. *Proc. of Wetland Soils Workshop, at IRRI*, pp. 191–206.

Mitsuchi, M. 1981. Characteristic features of paddy soils of Japan. *Proc. Symp. on Paddy Soils*, Nanjing, pp. 419–427.

Murphy, R.E. 1968. Landforms of the world. *Annals Assoc. Amer.*

Geographers, 58(1): Annals map supplement #9. (Cited from van Wambeke, A. 1992. *Soils of the Tropics, Properties and Appraisal.* McGraw-Hill, Inc., N.Y.).

Soil Survey Staff 1999. *Soil Taxonomy, 2nd Ed.: A Basic System of Soil Classification for Making and Interpreting Soil Surveys.* USDA Agric. Handb. No.436, U.S. Gov't. Printing Office, Washington, D.C.

Takaya, Y. 1968. Quaternary outcrops in the Central Plain of Thailand. In Takimoto, K. (Ed.), *Geology and Mineral Resources in Thailand and Malaya*, pp. 7–68. Center for Southeast Asian Studies, Kyoto University, Kyoto.

Takaya, Y. 1971. Physiography of rice land in the Chao Phraya Basin. *Tonan Ajia Kenkyu (Southeast Asian Studies)*, 9: 375–397.

Takaya, Y. and Tomosugi, T. 1972. Three categories of rice land in Northeast Thailand. *Ajia Keizai (Asian Economy)*, 13: 66–72. (In Japanese).

Thornthwaite, C.W. 1948. An approach toward a rational classification of climate. *The Geographical Review*, 38: 55–94.

White House 1967. *World Food Problem: A report of the President's Science Advisory Committee. Vol.II, Report of the Panel on the World Food Supply.* Washington, D.C.

Chapter 3: Chemical and Biological Changes of Paddy Soils in the Annual Cycle of Submergence and Drainage

3.1 Submerged period

It is essential for many paddy soils to have a natural or artificial cyclic process of submergence and drainage. Thus, a paddy soil experiences two entirely different chemical and biological processes in the course of a year—aerobic or oxidative and anaerobic or reductive. The aerobic or oxidative phase is the same as that in the upland condition, but the prolonged anaerobic or reductive phase is specific to the condition of paddy lands. The chemical changes in a soil caused by waterlogging or submergence are discussed first.

When a soil is submerged, free gas exchange between soil-air and atmosphere is strongly inhibited, and atmospheric oxygen can reach the soil only by diffusion through surface floodwater. The rate of oxygen diffusion through water is very slow (1/10000 of that in a gas phase), so the soil becomes anaerobic, as trapped and dissolved oxygen is consumed by the respiration of plant and soil organisms. The empirically observed changes that accompany the reduction process are described below for each of the solid, solution, gas and biotic phases, although they are naturally closely interrelated.

3.1.1 Solid phase

The first visible change for a submerged soil occurs in soil color. The brownish, yellowish or reddish tint of an oxidized soil turns to grayish, bluish or greenish within a few weeks after submergence, provided that adequate amounts of organic matter are present. The color change is due to the reduction of ferric iron and to the formation of ferrous iron. This is a kind of gleization phenomena, and is often denoted as surface water gleying or inverted gleying (or epiaquic saturation in Soil Taxonomy terms), when there is an oxidized subsurface or subsoil layer above the permanent ground water level (see Chapter 7.5).

Several weeks after submergence, brown coloring at the soil/water interface normally occurs, which is due to the differentiation of a thin, oxidized layer. Takai and Uehara (1973) studied this layer differentiation in the submerged plow layer of a paddy soil. In the initial period of submergence, the entire soil layer was reduced, accompanying a lowering of the redox potential and an increase of ferrous iron and NH_4-N. However, after this initial period, as NH_4-N formation ceased to be vigorous the amount of ferrous iron decreased and the redox potential increased markedly at the surface.

Patrick and Sturgis (1955) measured oxygen concentration in the surface water and in the different layers of soil, and showed that oxygen does not penetrate into soil layer by more than 1 cm. Patrick and Mikkelson (1971) proposed a schematic drawing of oxygen distribution across the soil/water interface, as shown in Figure 3.1. A very sharp transition from a thin oxidized layer to a strongly reduced layer is recognized.

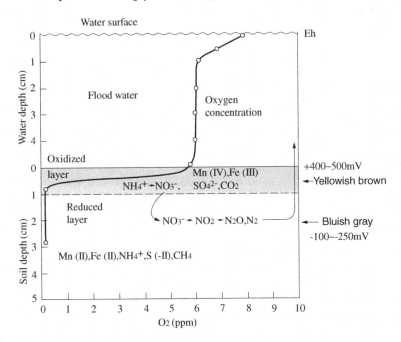

Figure 3.1
Pattern of oxygen distribution in paddy soil and forms of mineral constituents after stabilization (Source: Takai, 1978, modified after Patrick and Mikkelson, 1971)

These laboratory results suggest that differentiation of an oxidized layer at the soil/water interface is a result of oxygen dissolution in, and diffusion through, surface water. Some oxygen could be supplied even by the photosynthesis of aquatic plants. At the initial period of submergence, readily decomposable organic matter is still abundant, and oxygen reaching the soil/water interface is used up by vigorous microbial activities, so reducing the entire soil layer. However, as the rate of organic matter decomposition declines, the supply of oxygen exceeds its rate of consumption by microbes, thus resulting in the differentiation of an oxidized layer.

The thickness of the oxidized layer is determined by the balance between the rate of supply and the rate of consumption of oxygen. The brown coloring of the oxidized layer is due to the ferric oxides deposited by oxidation of ferrous iron diffused from the underlying reduced soil layer. The oxidized layer is the site where most of the reactions of aerobic metabolism by aerobic autotrophs occur. Of particular importance is nitrification, which is discussed later in relation to nitrogen fertility (see Chapter 8).

3.1.2 Soil/solution interface and solution phase

The substances produced or released during the reduction process are either adsorbed onto the soil surface or remain dissolved in soil solution. The greater part of NH_4^+ ions, liberated from decomposing organic matter, as well as Fe^{2+} and Mn^{2+} ions, as reduction products from ferric and manganic compounds, are adsorbed by soil particles, being exchanged mainly for Ca^{2+} and Mg^{2+}. However, they also appear in soil solution mainly with bicarbonate ions as anions, which increases the alkalinity of the soil solution. There have been many studies that tracked with different extractants, as a function of time, the changes in the amounts of these substances present at the soil/solution interface and in the solution phase of submerged soils. Some of the typical results are in Figure 3.2 (Takai *et al.*, 1969). The figure shows a rapid disappearance of O_2 and NO_3^- within a few days of submergence. Ferrous irons and CO_2 increased significantly during the first three weeks, and the former then levels off, while the latter declines sharply to its original level. The increase in acetic acid and sulfides was delayed slightly, and evolution of methane was delayed even further. Another experiment showed that NH_4-N follows a similar pattern to Fe^{2+}.

Such a sequential pattern of evolution and disappearance of the products under submergence has been noted in many studies and summarised by Patrick and Reddy (1978), as in Figure 3.3. The figure shows that the

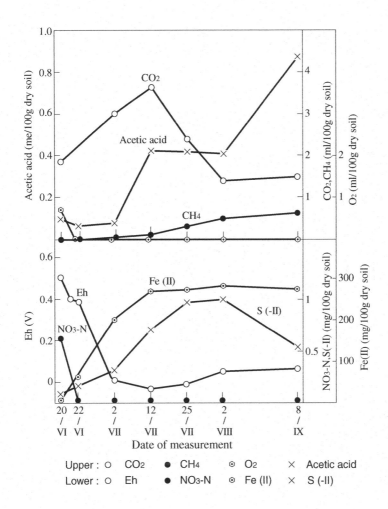

Figure 3.2
*Changes in Eh and various substances in submerged paddy soil under rice plants
in pot experiment (Source: Takai et al., 1969)*

sequential pattern is not necessarily followed strictly. For example, while oxygen is not completely consumed, NO_3^- is reduced and the reduction of Mn^{4+} to Mn^{2+} occurs while the reduction of O_2 and NO_3^- is still ongoing. However, the complete disappearance of NO_3^- lags behind that of O_2, and Fe^{3+} reduction does not occur until O_2 and NO_3^- have been almost completely depleted. Sulfides are formed only after the complete depletion of O_2 and NO_3^-.

Concentration (not to scale)

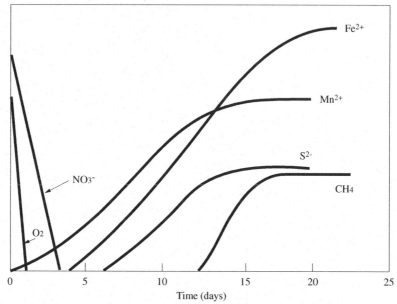

Figure 3.3
Sequential reduction of several soil redox systems in paddy soils (Source: Patrick and Reddy, 1978)

In strongly anaerobic conditions, organic acids are formed as intermediate products of organic matter decomposition. Acetic acid is the most commonly found volatile fatty acid in submerged soil solution. When readily decomposable organic matter is present in quantity, a toxic amount of butyric acid is formed in addition to acetic acid. Formic acid is another volatile fatty acid present in small quantities in submerged soil solution. Propionic, valeric, succinic and lactic acids are rarely detected. These organic acids are further decomposed to form CO_2 and/or CH_4 (as discussed in Chapter 5).

The pH of the soil solution of submerged acid soils is known to rise gradually as the reduction proceeds. Some examples are shown in Figure 3.4, in which the pH of soils rapidly increases as the redox potential (*Eh*) is lowered. When *Eh* levels off, the rise in pH stops. The soils with an alkaline pH prior to flooding, conversely, have their pH lowered due to the evolved CO_2 during the reduction process. Thus, the steady state pH after submergence often approaches neutral regardless of the original pH of the

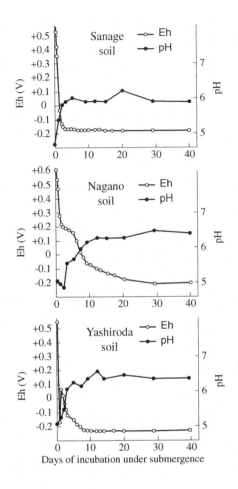

Figure 3.4
Eh *and pH of submerged soils (Source: Takai* et al.*, 1957)*

soil, as shown in Figure 3.5 (Ponnamperuma, 1972). (The redox potential will be dealt with in detail in Chapter 4.)

The carbonate content of the submerged soil solution is often very high. It has been reported that the total carbonate concentration in the soil solution is frequently as high as $10^{-1.5}$ to 10^{-2} M (Yamane, 1974). In terms of partial pressure of carbon dioxide gas with which the soil solution is in equilibrium, it is in the range of $10^{-0.7}$ (0.2) to $10^{-1.3}$ (0.05) atm (Ponnamperuma, 1972), which is much higher than $10^{-3.5}$ (0.0003) atm in the normal atmosphere.

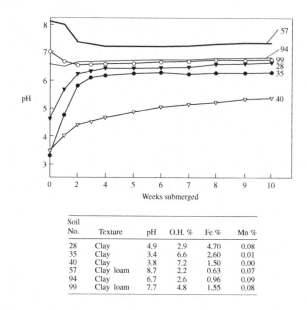

Soil No.	Texture	pH	O.H. %	Fe %	Mn %
28	Clay	4.9	2.9	4.70	0.08
35	Clay	3.4	6.6	2.60	0.01
40	Clay	3.8	7.2	1.50	0.00
57	Clay loam	8.7	2.2	0.63	0.07
94	Clay	6.7	2.6	0.96	0.09
99	Clay loam	7.7	4.8	1.55	0.08

Figure 3.5
Kinetics of the pH values of some submerged soils (Source: Ponnamperuma, 1972)

3.1.3 Gas phase

Yamane (1961) studied the evolution of gases in submerged soils and found that CO_2, CH_4, H_2 and N_2 are the main components. However, as CO_2 is readily soluble in water, it does not appear in the gas phase unless water is supersaturated or agitated. CH_4 is invariably produced when the soil becomes strongly reduced. The temperature and type of organic matter present greatly affect the gas composition. When glucose is added and the temperature is kept low, H_2 gas is produced in large quantities. However, when the temperature is high, it disappears rapidly by reducing other compounds. According to Yamane (1957), NO_3^- is reduced to N_2 almost quantitatively by denitrification in Japanese paddy soils.

3.1.4 Biotic phase (1): microorganisms

Ishizawa *et al.* (1956) studied the changes in microbial population in a lysimeter during the rice growing period for:
- surface water;
- the thin oxidized surface layer (0–0.5 cm);
- the rest of the plow layer.

In the surface water, the number of green algae increased to 10^5 ml^{-1}, but then decreased gradually. On the other hand, blue-green algae increased gradually and attained 10^3 ml^{-1} at the beginning of August (about a month after submergence), but subsequently disappeared. Aerobes ranged from 10^5 to 10^6 ml^{-1}, and tended to be more constant in number.

The thin oxidized surface layer seems to be most adequate for the multiplication of both green and blue-green algae. Here, blue-green algae attained almost the same number (10^6 g^{-1}) as green algae at the beginning of the submerged period, but decreased more rapidly than green algae. Aerobes fluctuated between 10^5 and 10^6, but tended to be lower in number in the oxidized layer than in surface water.

Within the reduced plow layer, bacteria predominated attaining 10^7 g^{-1} —about 10 times more than in the thin oxidized surface layer. Strictly anaerobic sulfate-reducers and aerobes were detected throughout the submergence period in the reduced plow layer.

Furusaka et al. (1969) studied changes in microbial populations in the course of the submergence period. Their results, as in Figure 3.6, show that by submergence, the number of fungi (moulds) and actinomycetes decreased sharply, while bacterial counts increased. The majority of the bacteria (60–80%) comprises facultative anaerobes. Among the aerobes, gram positives and gram negatives are present in almost equal numbers, in contrast to upland soils where gram positives predominate.

Among the obligate anaerobes only Clostridia are found at the level of 10^{4-5} g^{-1}. It is still not evident why obligate anaerobes do not predominate even under anaerobic paddy soil conditions.

Both the obligate and facultative anaerobes change their species composition during the submergence period. One example is shown for facultative anaerobes in Table 3.1 (Takeda and Furusaka, 1970).

Generally the number of bacteria increases with moisture, whereas the numbers of fungi and actinomycetes decrease. This may explain the predominance of bacteria in paddy soils. Another characteristic of paddy soils—that gram-negatives and gram-positives are balanced in number—is also related to the high moisture content. Since the gram-negative bacteria are more susceptible to low moisture, they are comparatively fewer in upland conditions, while in paddy soils their numbers increase.

Kanazawa et al. (1981) studied the seasonal changes of bacterial populations in relation to various fertilizer treatments. Figures 3.7 and 3.8 show the results for aerobes and anaerobes, respectively. Two main factors seem to control the seasonal changes of aerobe populations:

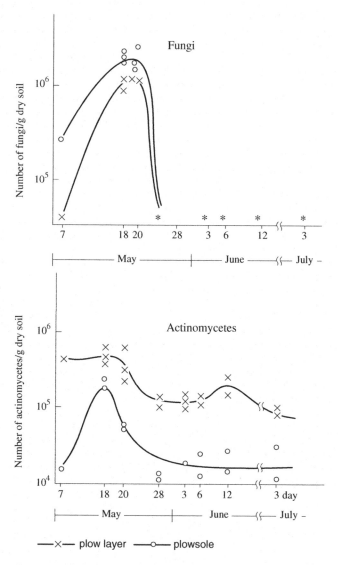

Figure 3.6
Fluctuation in the number of fungi and actinomycetes in the course of soil
submergence (Source: Furusaka et al., 1969)

- the redox condition of the soil that governs the changes for four months under submergence;
- the soil temperature during the eight months under the upland condition.

The latter is well demonstrated by the sharp decrease in bacterial count during the winter. No particular effect of fertilizer treatments was observed, except that higher counts were obtained throughout the year for the organic fertilizer and green manure plots as compared with no fertilizer and mineral fertilizer plots.

Anaerobes decrease their count under land preparation that aerates the plow layer. However, after submergence, a rapid increase occurs in every plot, particularly in the organic fertilizer plot. After draining, the number of anaerobes decreases and stays low during the winter. The relative difference between organic and green manure plots, on the one hand, and no fertilizer or mineral fertilizer plots, on the other, is even larger for anaerobes, although the absolute number of anaerobes is less than 10% that of aerobes.

Obligate aerobes dominate in the condition with >2 mm Hg partial pressure of oxygen, while at 0 mm Hg, obligate anaerobes dominate.

Table 3.1 Seasonal changes in the composition of facultative anaerobes

Species composition	Before submergence 15 May	During submergence 4 June	During submergence 12 Jul	After draining 9 Aug
Escherichia coli	52	–	–	–
Aerobacter aerogenes	4	–	5	–
Erwinia sp.	–	–	–	35
Aeromonas hydrophila	3	9	21	–
Streptococcus sp.	–	18	–	–
Staphylococcus sp.	–	–	3	–
Bacillus lichenformis	2	9	8	10
Bacillus pumilus	–	4	–	–
Bacillus cereus	–	2	–	–
Bacillus circulans	2	12	2	3
Bacillus polymyxa	–	1	–	–
F.V.	9	–	–	–
F.B.	–	2	3	1

Figures in the table indicate the number of strains isolated and identified at the date of sampling.
F.V.: Unidentified species that produce purple-colored pigments.
F.B.: Unidentified gram positive species.
(Source: Takeda and Furusaka, 1970)

Facultative anaerobes dominate when the O_2 partial pressure is within a narrow range of 1–2 mm Hg, which is the prevalent O_2 partial pressure condition in most paddy soils.

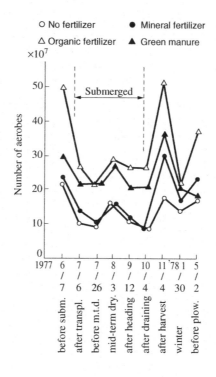

Figure 3.7
Seasonal fluctuation of bacterial count of aerobes in number per g dry soil at long-term fertilizer application plots (Source: Kanazawa et al., 1981)

3.1.5 Biotic phase (2): rhizosphere

Although rice is grown under submerged conditions, the rice roots have to respire to obtain energy for nutrient uptake. For this to occur, molecular oxygen or other terminal electron acceptors are necessary. As evidenced frequently, however, the concentration of oxygen in reduced soil is very low to nil. How then is it possible for the rice plants to grow?

The rice plant is known for its ability to supply molecular oxygen to its rhizosphere. Van Raalte (1941) first reported that rice plants could supply molecular oxygen to their roots. Mitsui *et al.* (1948) and Kumada (1948

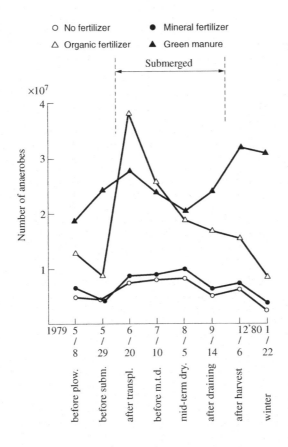

Figure 3.8
Seasonal fluctuation of bacterial count of anaerobes in number per g dry soil at long-term fertilizer application plots (Source: Kanazawa et al., unpublished, cited from Kanazawa, 1982)

a,b) also ascertained this. Mitsui *et al.* (1948) found that rice seedlings could transform leucomethylene blue in an agar-agar substrate medium to methylene blue by oxidation. Kumada (1948 a,b) used a special bacterium which emanates light in the presence of oxygen to prove that the roots of rice seedlings supply oxygen. They found that the oxidizing ability is localized to the tips of fresh white roots. Thus, rice can oxidize its rhizosphere, especially during the early growth stages through oxygen transmission from the atmosphere, suppressing sulfate reduction and sulfide formation.

It has been confirmed that the rice plant has oxidizing power, but at the same time it accelerates reduction of the rhizosphere at certain growth stages. Using the solution culture technique, Mitsui and Tensho (1951, 1952) found a noticeable formation of nitrite in the culture solution and attributed this mainly to the supply of reducing metabolic products from the top to the root, and also to a decrease in oxygen partial pressure because of root respiration. Furthermore, they found that the presence of the rice plant in a pot, increased organic matter in the leachate at the heading stage compared to an unplanted pot (Mitsui and Tensho, 1953).

Many other researchers found that ferrous iron in the soil solution or leachate is more abundant in planted plots than in unplanted plots (Yamanaka et al., 1956). Takai et al. (1969) and Kiuchi and Omukai (1959) attributed the increased ferrous iron leaching from the planted plot to:
- respiration by the roots of the rice plant;
- the supply of organic matter from root debris;
- the supply of organic matter by living roots as exudates.

Okajima (1958, 1960) and Shiga and Suzuki (1963) reported that when nitrogen is deficient, the rice rhizosphere tends to be more reductive.

The rice rhizosphere may be viewed as another aerobic-anaerobic interface in addition to the one formed between surface water and the soil. The aerobic environment of the rhizosphere may be confirmed by the presence of aerobic bacteria and nematodes on, or in, rice roots (Kawata et al., 1964), and Beggiatoa, a filamentous bacterium capable of oxidizing H_2S, colonizing on rice roots (Joshi and Hollis, 1977). However, autotrophic nitrification in the rice rhizosphere has not yet been confirmed. (Nitrogen fixation in the rice rhizosphere is discussed in Chapter 8.)

Asanuma et al. (1979) studied the bacterial flora in rhizosphere and non-rhizosphere soils using scanning electron microscopy. The rhizosphere had Pseudomonas and Bacillus, but their species composition was very simple. In non-rhizosphere soils, Bacillus predominated at more than 80% and was totally devoid of Pseudomonas that was dominant in the rhizosphere. However, the species composition was more diversified in the non-rhizosphere soils.

Kanazawa (1982) found that the rhizosphere effect (the ratio of the number of rhizosphere microorganisms to that in the non-rhizosphere) of rice, is exceptionally low compared with that of upland crops for any of the microbial groups. He attributed this to a rapid dilution of exudates from roots by submergence water, and to growth inhibition of aerobes due to the reductive condition of paddy soils. Anaerobes increase in the rhizosphere towards the latter stages of rice growth, accompanying an increase in the rhizosphere effect.

3.1.6 Subsoil condition

The rapid changes in oxidation and reduction conditions in the surface plow layer are reflected in the chemical and biological processes. Although these processes can also operate in the subsoil, their intensities are governed by a number of other factors. A major controlling factor is the ground water level. When the ground water level is high during the submergence period, the entire solum may be strongly reduced. However, when it is low, the subsoil shows relatively little redox change. Morita (1940) reported that in the double-cropped *kanden* (or dry paddy field, which is submerged for the cropping season and drained towards harvest) the subsoil redox potential (*Eh*) also tended to decline, but was considerably higher than the surface soil *Eh*.

Aomine (1949) measured the vertical *Eh* distribution for wet paddy fields (water saturated or submerged throughout a year) and dry paddy fields, and found that the subsoil of a dry paddy field always maintained *Eh* values higher than 30 mV. Yamazaki's results (1960) on ordinary dry paddy field soil, degraded *akiochi* soil (see Chapter 9), and wet paddy field soil with a permanently high water table, are shown in Figure 3.9. In a dry paddy field soil, only the surface soil shows a major decrease of *Eh*, while in a degraded *akiochi* soil that is low in free iron oxides the *Eh* drops even in the subsurface horizon. The *Eh* of a wet paddy field soil is low throughout the profile and throughout the period.

De Gee (1950) measured the *Eh* of a *sawah* soil profile in West Java and found that:

- Surface water *Eh* was 400–600 mV, but approximately 200 mV in the uppermost 2 cm layer of the plow layer;
- Between 10 and 16 cm from the soil surface, *Eh* was lowest, 0–100 mV;
- *Eh* rose sharply in the 16–20 cm layer, about 450 mV;
- Below 20 cm, *Eh* rose gradually and reached 600 mV at a 35 cm depth, which was identical to the non-submerged control soil.

Such a vertical *Eh* pattern has profound significance for the profile development of paddy soils, which is discussed in Chapter 7.

3.2 Drained period

Even during the rice-growing period, surface water is often drained for a short time, such as in the mid-term drying after the maximum tillering stage of rice, and in the intermittent irrigation practiced during the later period of the rice-cropping season. However, in many paddy soils under the typical monsoon climate, with distinct dry and wet seasons, or in those without

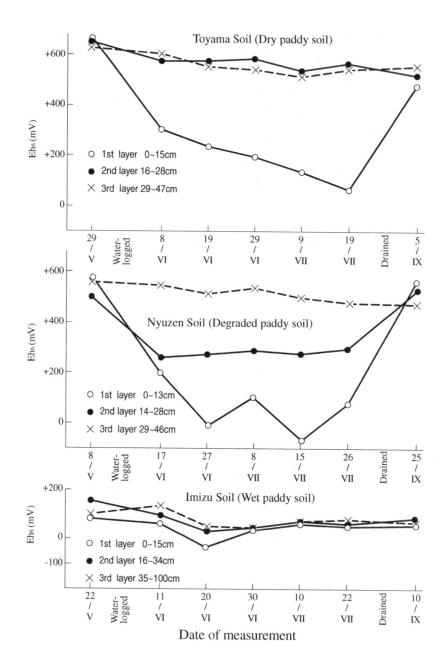

Figure 3.9
Changes in Eh of different layers of paddy soils before, during and after submerged
period (Source: Yamazaki, 1960)

perfect water control measures, the surface water is drained or dries up only towards the end of, or after, the growing season, and the soil is kept drained until the next rainy season.

During the drained period, soil behavior is generally not different from that in the upland condition, so it may be considered that there is nothing specific to mention. However, the changes in soil properties during the transitional period from wet to dry condition are of interest.

Some of the microbiological data for the drained period have been presented already, along with those for the submerged period.

3.2.1 *Eh* and pH changes upon draining

During the drained period, the chemical and biological processes are reversed to what has been reported in the preceding sections. Figure 3.9 shows the changes in the *Eh* in three paddy soils with groundwater at different levels. The return of the *Eh* to the original level seems to be a rapid process if drainage is effective. The pH changes are parallel to the *Eh* changes, and the surface soil pH of these three soils changed:

- from 5.6–6.3–5.5 for Toyama (dry paddy field) soil;
- from 5.7–5.8–5.8 for Nyuzen (degraded paddy) soil;
- from 5.8–6.5–6.1 for Imizu (wet paddy field) soil.

The Nyuzen soil is an *akiochi*-inclined soil with a low iron content, which may explain the small change in pH during the submergence period. It is expected that the final pH during the drained period eventually returns to the initial value.

3.2.2 Fe^{II} change upon drying

Motomura (1969) studied the changes in Fe^{II} of Nagano soils on drying and found 815 mg of extractable Fe^{II} per 100 g of dry soil by submergence. He fractionated the extractable Fe^{II} into the following four fractions:

- a water soluble fraction, which occurs as either an ionic or water soluble organo-mineral complex form and extracted with decarbonized water;
- an exchangeable fraction, which is exchangeably adsorbed by soil particles and extracted with N KCl at pH 7;
- an active fraction, which is insoluble precipitates of Fe^{II}, strongly adsorbed by soil particles and extracted with a 0.2% AlCl$_3$ solution or with pH 3.0 N Na-acetate;
- an inactive fraction, which is an insoluble precipitate of Fe^{II}, even more strongly adsorbed by soil particles and extracted only with a 0.2N HCl solution.

The results of his study are shown in Figure 3.10. There was a rapid decrease in extractable Fe^{II} within 48–72 hours. At the end of this period, the water content leveled off. However, after this stage, the decrease was very slow and even seven days after the start of drying, 305 mg of Fe^{II} (or about 35% of the maximum content of Fe^{II} in the reduced soil), remained in the soil.

Water-soluble Fe^{II} disappeared rapidly, and 34 hours after the start of drying, it constituted less than 5% of the water soluble Fe^{II} in the reduced soil. Exchangeable Fe^{II} also decreased rapidly, and 72 hours later, there remained less than 4% of the exchangeable Fe^{II} originally present in the reduced soil. Alternatively, active and inactive Fe^{II} were very persistent, and even after seven days, 49% and 77% of the respective amounts in the reduced soil were recovered.

It is clear that water-soluble and exchangeable Fe^{II} are very susceptible to oxidation by air, while active and inactive Fe^{II} are quite persistent. This finding parallels an observation in the field that soils that are high in water-soluble and exchangeable Fe^{II} in the reduced state turn rapidly from bluish gray or greenish gray to yellowish gray or yellowish brown upon drainage, whereas soils high in active and inactive Fe^{II} in the reduced condition, change only to grayish white after air-drying.

3.2.3 Mottle formation

In the surface plow layer, the water soluble and exchangeable Fe^{II} are oxidized rapidly and turn into different kinds of mottles. Filmy mottles are formed on the walls of crack surfaces, which are formed by dehydration after drainage. Ferrous ions diffuse from the interior of the soil matrix to cracks, and are deposited by air oxidation. Tubular mottles are mainly associated with rice roots. Oxygen diffuses into root channels after the roots are partially dehydrated or decayed. Here, the precipitation of ferric ions occurs in the same manner as in filmy mottles. The formation of spotty or concretionary mottles in the subsoil might be related to bacterial oxidation (iron bacteria) of ferrous ions. The mechanism of formation of the reticulated mottle pattern (or cloudy mottles) in the subsoil is difficult to explain. *In situ* segregation of iron in the cycle of oxidation and reduction might lead to the reticulated mottles.

The iron minerals contained in the mottles of paddy soils may be one or more of the following:

Amorphous hydrated ferric oxides	$Fe_2O_3.nH_2O$
Crystalline hydrated ferric oxides	
Goethite	α-FeOOH
Lepidocrocite	γ-FeOOH

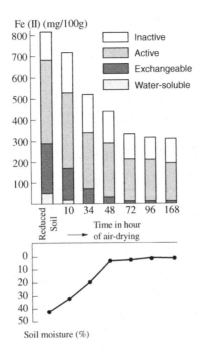

Figure 3.10
Changes in Fe^{II} Forms of Nagano soil during air-drying (Source: Motomura, 1969)

Crystalline ferric oxides
    Haematite α-Fe$_2$O$_3$
    Maghemite γ-Fe$_2$O$_3$
Other iron minerals
    Siderite   FeCO$_3$
    Vivianite   Fe$_3$(PO$_4$)$_2$.8H$_2$O
    Jarosite   KFe$_3$(SO$_4$)$_2$(OH)$_6$

Of these minerals, crystalline ferric oxides, hematite and maghemite, are rare and usually not found in paddy soils. According to Kojima and Kawaguchi (1968), goethite is common in paddy soils in tropical Asia and lepidocrocite is common in paddy soils of Japan. Siderite and vivianite occur only in strongly reduced paddy soils with a high amount of organic matter, such as soils with an underlying or interbedded peat layer. Jarosite forms straw-yellow colored brittle mottles, and is indicative of acid sulfate soils or cat-clays (see Chapter 11). It represents an intermediate product of oxidation of pyrite in an acidic medium.

Manganese (Mn) is also accumulated as mottles, especially in the subsurface and subsoil. Because it is so readily reduced and mobilized in the reduction process and because it is much lower in content, Mn mottles in the surface plow layer are rarely observed. Filmy, tubular, and spotty mottles of manganese are known, and they are formed in the same manner as corresponding iron mottles. Here too, a bacterial origin for spotty Mn mottles is highly probable.

3.2.4 Exchangeable cation composition

In the reduced phase of paddy soils, a large amount of Fe^{II} is formed and ferrous ions replace some of the exchangeable cations originally present in the soil. In Motomura's experiment (see Figure 3.10), about 200 mg or some 8 me of Fe^{II} per 100 g of soil were present as exchangeable ions in the Nagano soil (LiC, montmorillonitic). Upon drainage, this amount of exchangeable Fe^{II} disappeared within a few days and the exchange site which Fe^{II} had occupied must have been taken over by some other cations. Thus, the change in exchangeable cation composition is inevitable.

Kawaguchi and Kawachi (1969a) showed that Fe^{II} is adsorbed strongly by the exchange complex in the reduced condition and replaces exchangeable calcium (Ca) by as much as 50% or more. They leached Ca-saturated soil columns with water, one in an oxidized condition and the other in a reduced condition, by the addition of glucose. Exchangeable cation compositions, after the completion of the leaching experiment, are shown in Table 3.2. The column kept in the reduced condition shows, a higher degree of leaching loss of exchangeable Ca, a higher amount of exchangeable NH_4, and higher contents of exchangeable H and Al. At the same time, the CEC of the soil in this column is lowered appreciably.

Table 3.2 Exchangeable cation composition of previously Ca-saturated soils after a leaching experiment

Treatment	Soil	pH Before experiment	pH After experiment	Exchangeable cations ($cmol\ kg^{-1}$) Ca	NH_4	H	Al	Sum	Leaching loss of Ca %
Oxidative	A	5.6	5.4	11.2	2.52	0.32	0.10	14.1	27.3
	B	4.9	5.0	17.3	2.82	1.26	0.12	21.5	16.4
	C	5.1	5.2	5.4	1.46	0.23	0.10	7.2	36.5
Reductive	A	5.6	5.0	6.7	4.04	0.62	0.21	11.6	56.5
	B	4.9	4.5	12.9	4.62	1.32	0.60	19.4	37.7
	C	5.1	4.0	2.1	1.36	1.78	0.87	6.1	75.3

* Ca saturation was done in an unbuffered medium.
(Source: Kawaguchi and Kawachi, 1969a)

Kawaguchi and Kawachi (1969b) related the acidification and increase of exchangeable H and Al to ferrous ions, which occupied exchange sites during the submerged period and later disappeared upon drying and oxidation of the soil. They postulated the following reaction mechanism:

$$Fe^{2+} - clay \xrightarrow{H_2O} H^+ - clay + Fe(OH)_2$$

$$Fe(OH)_2 \xrightarrow{H_2O} Fe(OH)_3 + H^+$$

This reaction is better written as follows:

$$Fe^{2+}\text{-clay} + (1/4)O_2 + (5/2)H_2O \rightarrow (H^+)_2\text{-clay} + Fe(OH)_3$$

Part of the H^+-clay so formed is further spontaneously transformed to Al^{3+}-clay, thereby resulting in the partial destruction of the clay lattice, which in turn may explain the decrease in the CEC. Kawaguchi and Kawachi (1969b) also measured the amounts of exchangeable Al and silica released in the course of the drying of Fe^{2+}-montmorillonite. Depending on the partial pressure of oxygen during the drying process, the extent of clay destruction varied, that is, the higher the oxygen partial pressure, the higher the amounts of Al and SiO_2 released.

Yoshida and Itoh (1974) confirmed the above process both in the field and in the laboratory, as shown in Table 3.3. They concluded that:

Table 3.3 Composition of exchangeable cations in the soils under reduced condition and after oxidation

Soil and its redox state		Exchangeable cations (cmol kg⁻¹ dry soil)									PBS[a] %	PFS[b] %	Active Fe[II] cmol kg⁻¹
		Ca	Mg	K	Na	NH₄	Fe[II]	Al	H	Sum			
Soil A	Red.	13.1	2.7	0.5	0.2	0.2	7.8	0.0	9.0	33.6	73	23	9.9
	Ox.	11.6	2.8	0.3	0.2	0.6	0.0	0.0	17.9	33.3	46	0	0.3
Soil B	Original	11.6	1.1	0.4	0.3	0.3	0.0	0.1	8.7	22.5	61	0	0.0
	Red.	9.6	2.7	1.6	0.2	1.6	5.1	0.0	2.2	23.0	90	22	6.3
	Ox.	10.3	1.0	0.3	0.2	1.8	0.0	0.2	8.4	22.2	61	0	0.0
Soil C	Original	9.8	1.1	0.4	0.3	0.2	0.0	1.2	8.6	21.6	57	0	0.0
	Red.	10.3	2.2	0.6	0.5	1.0	7.4	0.0	2.8	24.8	89	30	10.3
	Ox.	9.2	2.1	0.9	0.2	1.1	0.2	1.4	6.9	22.0	63	1	0.3

Notes:
a) Percentage base saturation
b) Percentage Fe[II] saturation
(Source: Yoshida and Itoh, 1974)

• The degree of acidification is determined primarily by the degree of Fe^{2+} saturation in the exchange complex, which ranged from 22 to 30% for the three samples they studied;

• The occurrence of Fe^{2+}-H^+-Al^{3+} replacement and clay destruction depends on the acid strength of the exchange sites. One soil with large amounts of allophane and organic matter showed only Fe^{2+}-H^+ replacement and no exchangeable Al^{3+} appeared upon drying.

This phenomenon of acidification, and the clay destruction of paddy soils in their submergence-draining cycle, has been related closely to the process that was later termed 'ferrolysis' by Brinkman (1970), for seasonally submerged soils. This topic is discussed further in Chapter 7, which deals with long-term changes in the morphology and properties of paddy soils.

References

Aomine, S. 1949. *Tile drainage and soil-drying effect*. Kawade-shobo. (Cited from Kawaguchi, K. (Ed.) 1978. *Paddy Soil Science*. Kodansha, Tokyo. In Japanese).

Asanuma, S., Tanaka, H. and Yatazawa, M. 1979. Rhizoplane microorganisms of rice seedlings as examined by scanning electron microscopy. *Soil Sci. Plant Nutr.*, 25: 539–551.

Brinkman, R. 1970. Ferrolysis, a hydromorphic soil forming process. *Geoderma*, 3: 199–206.

De Gee, J.C. 1950. Preliminary oxidation potential determinations in a 'Sawah' profile near Bogor (Java). *Trans. 4th Int'l Congr. Soil Sci., Amsterdam*, 1: 300–303.

Furusaka, C., Hattori, T., Sato, K. Yamagishi, H., Hattori, R., Nioh, T. and Nishio, M. 1969. Microbiological, chemical and physicochemical surveys of the paddy field soil. *Rep. Inst. Agric. Res. Tohoku Univ.*, 20: 89–101.

Ishizawa, S., Toyoda, K. and Tanabe, I. 1956. *Outlines of the experimental results in 1955*. National Institute of Agricultural Sciences, Tokyo. (In Japanese).

Joshi, M.M. and Hollis, J.P. 1977. Interaction of *Beggiatoa* and rice plant: detoxification of hydrogen sulfide in the rice rhizosphere. *Sci.* 195: 179–180.

Kanazawa, S. 1982. Chapter 6. Biota in paddy soils. In Yamane, I. (Ed.) *Paddy Soil Science*, pp. 233–279. No-Bun-Kyo, Tokyo. (In Japanese).

Kanazawa, S., Hasebe, R. and Takai, Y. 1981. Microbial number and activity in paddy fields under a long-term continuous fertilizer application trial. (2) Seasonal changes of microbial number in plow layer. (Cited from Kanazawa, 1982) (In Japanese).

Kawaguchi, K. and Kawachi, T. 1969 a. The cation exchange reactions in the submerged soil. *J. Sci. Soil & Manure, Japan*, 40: 89–95. (In Japanese).

Kawaguchi, K. and Kawachi, T. 1969 b. Changes in the composition of exchangeable cations in the course of drying of preliminarily submerged soils. *J. Sci. Soil & Manure, Japan*, 40: 177–183. (In Japanese).

Kawata, S., Ishihara, K. and Iizuka, H. 1964. Several microorganisms on or in rice roots. *Proc. Crop Sci. Soc. Jpn.*, 33: 164–167. (In Japanese).

Kiuchi, T. and Omukai, S. 1959. Influence of organic matter content and plant root on the leaching of cation from paddy soils. *Soil & Plant Food*, 5: 108–113.

Kojima, M. and Kawaguchi, K. 1968. Identification of free iron minerals in rusty mottles in paddy soils in Japan. *J. Sci. Soil & Manure, Japan*, 39: 349–353. (In Japanese).

Kumada, K. 1948 a. Investigation on the rhizosphere of rice seedling (Part 1). On the microscopic structure of the rhizosphere and oxidative power of root. *J. Sci. Soil & Manure, Japan*, 19: 119–124. (In Japanese).

Kumada, K. 1948 b. Investigation on the rhizosphere of rice seedling (Part 2). On the oxygen consuming power of paddy soils. *J. Sci. Soil & Manure, Japan*, 19: 124–128. (In Japanese).

Mitsui, S. and Tensho, K. 1951. Dynamic studies on the nutrients uptake by crop plants (Part 3). The reducing power of the roots of growing plants as revealed by nitrite formation in the nutrient solution. *J. Sci. Soil & Manure, Japan*, 22: 301–307. (In Japanese).

Mitsui, S. and Tensho, K. 1952. Dynamic studies on the nutrients uptake by crop plants (Part 4). The mechanism of nitrite formation by metabolizing plant roots. *J. Sci. Soil & Manure, Japan*, 23: 5–8. (In Japanese).

Mitsui, S. and Tensho, K. 1953. Dynamic studies on the nutrients uptake by crop plants (Part 8). The reducing power of the root of growing rice plant as revealed by the soil leachate. *J. Sci. Soil & Manure, Japan*, 24: 145–148. (In Japanese).

Mitsui, S., Hashimoto, H. and Terasawa, S. 1948. Mechanism of physiological disorders of rice roots in degraded paddy soils. *J. Sci. Soil & Manure, Japan*, 19: 59–61 (Abstract). (In Japanese).

Morita, S. 1940. Studies on the oxidation-reduction potentials of soils (II). On the oxidation-reduction potential of soils under paddy and dry field condition. *J. Sci. Soil & Manure, Japan*, 14: 43–62. (In Japanese).

Motomura, S. 1969. Behavior and roles of ferrous iron in paddy soils. *Bull. Nat. Inst. Agric. Sci.*, B21: 1–114. (In Japanese).

Okajima, H. 1958. On the relationship between tha nitrogen deficiency of the rice plant roots and the reduction of the medium. *J. Sci. Soil & Manure, Japan*, 29: 175–180. (In Japanese).

Okajima, H. 1960. Physiological functions of rice roots with special emphasis on nitrogen nutrition. *Bull. Inst. Agric. Res. Tohoku Univ.*, 12: 1–146. (In Japanese).

Patrick, W.H. and Reddy, C.N. 1978. Chemical changes in rice soils. In IRRI (Ed.) *Soils and Rice*, pp.361–379. IRRI, Los Baños, Philippines.

Patrick, W.H. and Mikkelsen, D.S. 1971. Plant nutrient behavior in flooded soil. *Fertilizer Technology and Use*, 2nd Ed., Soil Sci. Soc. Am., Inc., Madison, Wisconsin, pp. 187–215.

Patrick, W.H. and Sturgis, M.B. 1955. Concentration and movement of oxygen as related to absorption of ammonium and nitrate nitrogen by rice. *Soil Sci. Soc. Amer. Proc.*, 19: 59–62.

Ponnamperuma, F.N. 1972. The chemistry of submerged soils. *Adv. in Agron.*, 24: 29–96.

Shiga, H. and Suzuki, S. 1963. Behavior of hydrogen sulfide in flooded soils, Part 6. Iron that reacts with hydrogen sulfide evolved in flooded soils. *Bull. Chugoku Nat. Agric. Expt. Sta.*, A7: 197–220. (In Japanese).

Takai, Y. 1978. Redox processes in the soil under submergence. In Kawaguchi, K. (Ed.) *Paddy Soil Science*, pp.23–55. Kodan-sha, Tokyo. (In Japanese).

Takai, Y. and Uehara, Y. 1973. Nitrification and denitrification in the surface layer of submerged soil (Part 1). Oxidation-reduction condition, nitrogen transformation and bacterial flora in the surface and deeper layers of submerged soils. *J. Sci. Soil & Manure, Japan*, 44: 463–470. (In Japanese).

Takai, Y., Koyama, T. and Kamura, T. 1957. Microbial metabolism of paddy soils, Part III. Effect of iron and organic matter on the reduction process (1). *J. Agric. Chem. Soc. Japan*, 31: 211–215. (In Japanese).

Takai, Y., Koyama, T. and Kamura, T. 1957. Microbial metabolism of paddy soils, Part IV. Effect of iron and organic matter on the reduction process (2). *J. Agric. Chem. Soc. Japan*, 31: 215–220 (In Japanese).

Takai, Y., Koyama, T. and Kamura, T. 1969. Effects of rice plant roots and percolating water on the reduction process of flooded paddy soil in pot. Microbial metabolism in paddy soils (Part 5). *J. Sci. Soil & Manure, Japan*, 40: 15–19. (In Japanese).

Takeda, K. and Furusaka, C. 1970. On the bacteria isolated anaerobically from paddy field soil, Part I. *J. Agric. Chem. Soc. Japan*, 44: 343–348. (In Japanese).

Yamane, I. 1957. Nitrate reduction and denitrification in flooded soils. *Soil & Plant Food*, 3: 100–103.

Yamane, I. 1961. Method of gas analysis and the changes of gaseous components in paddy soils. *Bull. Inst. Agric. Res. Tohoku Univ.*, 12: 261–367. (In Japanese).

Yamane, I. 1974. Solution pH of submerged soils. *Rep. Inst. Agric. Res. Tohoku Univ.*, 25: 1–11.

Van Raalte, M.H. 1941 On the oxygen supply of rice roots. *Ann. Bot. Gardens, Buitenzorg*, 51:43–58. (Cited from Kawaguchi, K. (Ed.) 1978. *Paddy Soil Science*. Kodan-sha, Tokyo. In Japanese).

Yamanaka, K., Nakamura, H., Matsuo, K. and Motomura, S. 1956. Studies on soil gleization. *Abstracts of the 1956 Meeting, Soc. Sci. Soil & Manure, Japan*, 2: 6. (In Japanese).

Yamazaki, K. 1960. Genetic classification of paddy soils. *Rep. Toyama Agric. Expt. Sta.*, Spec. Issue 1: 1–105. (In Japanese).

Yoshida,M. and Itoh, N. 1974. Composition of exchangeable cations in submerged paddy soil and acidifying process associated with exposure to air. *J. Sci. Soil & Manure, Japan*, 45: 525–528. (In Japanese).

Chapter 4: Fundamental Chemical Reactions in Submerged Paddy Soils

4.1 The basic concept of a redox system

When soil is submerged by water, free gas exchange between soil-air and the atmosphere is strongly inhibited, as atmospheric oxygen can only diffuse through the water to the soil very slowly (about 1/10000 of that in a gas phase). Thus, when oxygen that is dissolved and trapped in a soil solution and a soil matrix has been consumed by plant and soil organisms, the soil becomes anaerobic or reduced. Under this state, some of the chemical species such as Fe^{III}, Mn^{IV} and NO_3^- are reduced to Fe^{II}, Mn^{II} and N_2 or N_2O respectively. The transformation of these chemical species with two or more different valence states is called an 'oxidation-reduction reaction'. In this chapter, the chemical aspects of the processes taking place in paddy soils are discussed while biological and biochemical aspects will be discussed in Chapter 5.

4.1.1 The redox potential

Oxidation-reduction is a reaction in which a transfer of electrons from a donor to an acceptor is involved. An electron donor is oxidized by losing electrons, and an electron acceptor is reduced by gaining electrons. Thus, oxidation and reduction reactions occur simultaneously, so are often called a 'redox reaction'. A system in which coupled redox reactions occur is called a redox system.

For a generalized redox half reaction of the following type:

$$Ox + ne \rightleftharpoons Red \qquad\qquad K \text{ (equilibrium constant)} \qquad (1)$$

the free energy change accompanying the reaction, ΔG, is expressed as:

$$\Delta G = \Delta G^\circ + RT \ln \frac{(Red)}{(Ox)} \qquad (2)$$

where ΔG° is the standard free energy change of the reaction, in which all of the components are at unit activity. (Ox) and (Red) are the respective activities of oxidized and reduced species, of a substance that undergoes the redox reaction, and 'n' is the number of moles of electrons transferred in the reaction.

In an electrochemical reaction in a reversible cell:

$$\Delta G = -nFE \tag{3}$$

as there is no change in temperature, pressure and volume. Here, E is the eletromotive force or electric potential for the reversible cell in V, F is the Faraday constant or 96490 C mol^{-1}, thus, nF is the electric charge required to carry out the reaction. When all the reactants and products are in the standard state, ΔG° is written as:

$$\Delta G^\circ = -nFE^\circ \tag{4}$$

where E° is the standard electric potential for the reaction. If either the E° or ΔG° value is known, the other can be calculated (1 volt-coulomb = 1 joule):

$$\Delta G^\circ = -96490 \, n \, E^\circ \text{ (J)},$$
or $\qquad E^\circ = -\Delta G^\circ / \, (96490 \, n) \text{ (V) at } 25°C.$

Substituting (3) and (4) for ΔG and ΔG° in (2), the following is obtained:

$$E = E^\circ + \frac{RT}{nF} \ln \frac{(Ox)}{(Red)} \tag{5}$$

This is the conventional expression of the Nernst equation for the reaction (1). When E is measured against the standard hydrogen electrode, it is written as Eh (or E_H) and called the 'redox potential'. Therefore:

$$Eh = E^\circ + \frac{RT}{nF} \ln \frac{(Ox)}{(Red)} \tag{6}$$

According to the International Union of Pure and Applied Chemistry (IUPAC) convention, E° is positive when log K (equilibrium constant) for the reaction (1) is positive.

Most of the redox reactions that occur in the natural environment involve hydrogen ions, H^+. For a redox reaction that involves mH^+, that is:

$$Ox + mH^+ + ne \rightleftharpoons Red \qquad K \text{ (equilibrium constant)} \tag{7}$$

The derived equation for the redox potential is:

$$Eh = E° + \frac{RT}{nF} \ln \frac{(Ox)(H^+)^m}{(Red)} \qquad (8)$$

As $\ln X = 2.3 \log X$ and $pH = -\log (H^+)$, equation (8) can be rewritten as:

$$Eh = E° + \frac{2.3RT}{nF} \log \frac{(Ox)}{(Red)} - \frac{2.3mRT}{nF} pH \qquad (9)$$

Substituting numerical values for R (8.314 J mol^{-1} K^{-1}), F (96490 C mol^{-1}) and T (298 K or 25°C), $2.3RT/F$ becomes 0.059 V at 25°C. Thus, the redox potential for the reaction (7) is written as:

$$Eh = E° + \frac{0.059}{n} \log \frac{(Ox)}{(Red)} - \frac{0.059m}{n} pH \qquad (10)$$

4.1.2 The concept of pE

Sillen (1967), a geochemist, proposed the concept of pE as a replacement for Eh. When the electron in the reaction (7) is treated as if it is a component of an ordinary chemical reaction, the new equilibrium constant, K', may be written as:

$$K' = \frac{(Red)}{(Ox)(H^+)^m(e)^n} = K \frac{1}{(e)^n} \qquad (11)$$

As pE is defined to be the negative logarithm of the electron activity, that is:

$$pE = -\log (e) \qquad (12)$$

from (11)

$$\begin{aligned} \log K' &= \log K - n \log (e) \\ &= \log K + n\ pE \end{aligned}$$

The standard free energy change of the reaction (7) is expressed as:

$$\begin{aligned} \Delta G° &= -nFE° \\ &= -RT\ln K' \end{aligned}$$

Thus:

$$E^\circ = \frac{RT}{nF} \ln K' = \frac{0.059}{n} \log K'$$

$$= -\frac{0.059}{n} \log \frac{(Ox)(H^+)^m}{(Red)} + 0.059pE \tag{13}$$

When (13) is substituted for E° in (8):

$$Eh = 0.059pE \tag{14}$$

In (14), when the activity of all the chemical species is in unity:

$$E^\circ = 0.059pE^\circ \tag{15}$$

Substituting (14) and (15) for Eh and E° in (10), (10) is rewritten as:

$$pE = pE^\circ + \frac{1}{n} \log \frac{(Ox)}{(Red)} - \frac{m}{n} pH \tag{16}$$

In strongly oxidized systems, the electron activity is low, so the pE is high and positive, whereas in strongly reduced systems, the pE is low or even negative.

It should be noted that the redox potential Eh or pE values are rather insensitive to the activity or concentration changes of (Ox) or (Red). When there is a ten times difference in the ratio of the two, it will produce only $59/n$ mV change in Eh, or $1/n$ unit change in the pE scale. Thus, the general level of Eh is determined by a system that has the standard redox potential in the vicinity of the measured Eh value.

4.2 Measurement of the redox potential

The redox potential is measured for an electrochemical cell set between an inert electrode, such as a Pt electrode, placed in a solution or a suspension with a redox system, and a reference electrode, usually a saturated calomel electrode, as follows:

Hg, Hg_2Cl_2 | KCl || Ox, Red | Pt

In this notation of an electrochemical cell, a single vertical line represents a phase boundary across which there is a potential difference, and a double vertical line signifies that the liquid junction potential[1] is either ignored or considered to be eliminated by a salt bridge. In the above notation, the electrode reaction on the right is a reduction (this electrode being a cathode), and the electrode reaction on the left is an oxidation (an anode).

Figure 4.1 illustrates an experimental setup for Eh measurement. Normally a pH meter is used as a potentiometer. While making a measurement, care must be taken not to allow the sample to be exposed to the air. In the field, soil sampling operation (putting a soil sample into a container and inserting a rubber stopper) is performed under water to prevent air entering the container.

The potential values of the calomel electrode as the reference electrode are as follows[2]:

Temperature	Calomel	
°C	0.1 M KCl	saturated KCl
12	0.3362 V	0.2528 V
20	0.3360 V	0.2508 V
25	0.3356 V	0.2444 V

These values include the liquid junction potential for the KCl bridge. An appropriate potential value of the calomel electrode must be added to the direct reading of the potentiometer for the cell to obtain the potential of the redox system in question, as $E_{cell} = Eh_{(Ox, Red)} - E_{ref}$.

A Pt electrode can easily be made by fusing a Pt wire of 0.5 mm diameter to a glass tube, leaving some of the wire outside. A small amount of Hg is placed in the glass tube, in which the Cu wire connecting the electrode to a potentiometer is immersed.

Ponnamperuma (1972) used a 10 mm long Pt wire, whereas Yamane and Sato (1968) used an 80 mm long coiled wire to create a sufficiently large surface area. They considered that a Pt electrode should have a surface area larger than 100 mm^2 to obtain reliable data. As it is difficult to handle such a long wire electrode, it is necessary to compromise on the length of the Pt wire. It would be more practical to make the wire length 10–20 mm, or to use a small Pt plate (5 × 10 mm) electrode welded to the Pt wire.

The Pt electrodes should be checked against a standard redox buffer solution, such as a solution of 0.0033 M K$_3$Fe(CN)$_6$ and 0.0033 M K$_4$Fe(CN)$_6$ in 0.1 M KCl with an Eh of 0.430 V at 25°C (Zobell, 1946) and a suspension of pure quinhydrone in 0.05 M potassium acid phthalate solution with an Eh of 0.463 V at 25°C (Ponnamperuma, 1972).

The electrodes must be cleaned thoroughly with a detergent solution, 1 N HCl and distilled water, and then ignited in an ethanol flame until red-hot.

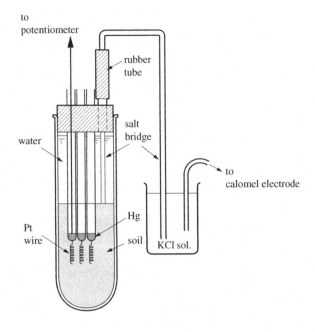

Figure 4.1
Eh *Measuring setup with three electrodes for triplicate measurements*
(Source: Yamane, 1961)

The Pt electrode surface is easily contaminated by colloidal materials in the soil and responds inaccurately to the redox potential change. This is one of the most difficult problems to overcome when continuously monitoring soil or solution *Eh* in the field by leaving a Pt electrode buried in the soil.

In spite of these precautions, *Eh* measurements of reduced soils have many limitations. First, the soil system is highly heterogeneous and it is difficult to attain a true equilibrium potential. Moreover, many of the important redox pairs, such as NO_3^-/NH_4^+, SO_4^{2-}/S^{2-}, CO_2/CH_4, and organic redox couples, are not electroactive, but they can interfere with the *Eh* measurements by producing 'mixed potentials'. According to Stumm and Morgan (1970), in natural water, accurate measurements of the *Eh* for the respective redox systems can be made only for Fe^{3+}/Fe^{2+} and Mn^{4+}/Mn^{2+} pairs existing in concentrations higher than 10^{-5} *M*. Thus, the *Eh* measured for reduced soils is almost always a mixed potential and is difficult to use in quantitative thermodynamic treatments.

Also, pH directly affects *Eh* or pE, as was shown in equations (10) and (16). As the redox system operating in the soil is not exactly known, it is impossible to make corrections of Eh for pH. Therefore, should be presented both uncorrected Eh and pH data instead of correcting the measured values to Eh at pH 7 by adopting the widely used correction factor of $dEh/dpH = -59$ mV, which is applicable only to a system with the ratio of $m/n = 1$, as shown in equation (10). Experimental values of the dEh/dpH range from some -60 mV to over -200 mV. In addition, there is a divergence of potentials measured for soil from those measured for soil solutions. Solutions rarely have a value lower than 0 V, while the soil potentials drop to as low as -0.3 V.

4.3 Redox systems operating in paddy soils

The ease with which any redox reaction proceeds can be judged from its standard redox potential, $E°$. A system with a high and positive $E°$ has a higher affinity for electrons and thus can be easily reduced, while a system with a low or negative $E°$ readily releases electrons to shift to an oxidized state. Therefore, if two redox pairs with different $E°$ values are connected, the one with a higher $E°$ is reduced and the one with a lower $E°$ is oxidized. Such a coupled redox reaction yields energy and the more the $E°$ differs, the more energy is generated. Important redox pairs operating in paddy soils under submerged conditions are listed in Table 4.1.

Table 4.1   Standard redox potential $E°$, and $pE°$ values for some redox systems important in submerged soils

Redox half reactions	$E°(V)$	$E°_7(V)$	$pE°$	$pE°_7$
$O_2(g) + 4H^+ + 4e = 2H_2O$	1.23	0.82	20.8	13.8
$(1/5)NO_3^- + (6/5)H^+ + e = (1/10)N_2 + (3/5)H_2O$	1.24	0.74	21.0	12.6
$(1/2)NO_3^- + H^+ + e = (1/2)NO_2^- + (1/2)H_2O$	0.83	0.42	14.1	7.1
$(1/8)NO_3^- + (5/4)H^+ + e = (1/8)NH_4^+ + (3/8)H_2O$	0.88	0.36	14.9	6.2
$*(1/2)MnO_2(s) + (1/2)HCO_3^- + (3/2)H^+ + e = (1/2)MnCO_3(s) + H_2O$	1.12	0.50	19.0	8.5
$(1/2)MnO_2(s) + 2H^+ + e = (1/2)Mn^{2+} + H_2O$	1.23	0.40	20.8	6.8
$*FeOOH(s) + HCO_3^- + 2H^+ + e = FeCO_3(s) + 2H_2O$	0.73	-0.096	12.3	-1.67
$Fe(OH)_3(s) + 3H^+ + e = Fe_2^+ + 3H_2O$	1.06	-0.18	17.9	-3.13
$(1/2)CH_2O + H^+ + e = (1/2)CH_3OH$	0.24	-0.178	4.0	-3.01
$(1/8)SO_4^{2-} + (9/8)H^+ + e = (1/8)HS^- + (1/2)H_2O$	0.24	-0.221	4.1	-3.75
$(1/8)CO_2(g) + H^+ + e = (1/8)CH_4(g) + (1/4)H_2O$	0.16	-0.244	2.9	-4.14
$(1/6)N_2(g) + (4/3)H^+ + e = (1/3)NH_4^+$	0.27	-0.277	4.6	-4.69
$H^+ + e = (1/2)H_2(g)$	0.00	-0.413	0.0	-7.00
$(1/4)CO_2 + H^+ + e = (1/4)CH_2O + (1/4)H_2O$	-0.071	-0.484	-1.2	-8.2

* HCO_3 concentration in the equation is set at $10^{-3}M$
(Sources: Stumm and Morgan, 1970; Ponnamperuma, 1972)

4.3.1 Redox reactions in natural water

In nature, water is the medium for most chemical reactions. Water is made of hydrogen and oxygen. As oxygen is reduced to water, the redox reaction for the O_2-H_2O pair can be written as:

$$O_2 + 4H^+ + 4e = 2H_2O \tag{17}$$

As shown in Table 4.1 $E°$ for (17) is 1.23 V or $pE° = 20.80$, indicating that oxygen is a very strong oxidizing agent. If an even stronger oxidizing system with a higher $E°$ than that of the oxygen system is put into water, the above reaction proceeds backwards and water is oxidized to liberate oxygen gas. For example, if Cl_2 gas is injected into water, oxygen is produced and HCl is formed, as shown below:

$$\begin{array}{ll} Cl_2 + 2e = 2Cl^- & (E° = 1.36\ V) \\ \underline{H_2O = (1/2)\ O_2 + 2H^+ + 2e} & (E° = 1.23\ V) \\ Cl_2 + H_2O = 2H^+ + 2Cl^- + (1/2)\ O_2 & \end{array}$$

In nature, such a reaction leading to the decomposition of water rarely occurs, so the oxygen system (17) can be regarded as the strongest oxidizing system.

On the other hand, water is dissociated into H^+ and OH^-. The H^+ ions represent an oxidized state relative to hydrogen gas, H_2. The hydrogen system may be written as:

$$2H^+(aq) + 2e = H_2(g) \tag{18}$$

At pH = 0, the $E°$ of the hydrogen system is, by definition, zero at its standard state. The metal Na is a strong reducing agent ($Na^+ + e = Na(s)$: $E° = -2.71$ V) in the sense that it is so readily oxidized in nature. If metallic Na is placed in water, the overall reaction is:

$$Na(s) + H_2O = Na^+ + OH^- + (1/2)\ H_2$$

resulting in the decomposition of water to liberate H_2 gas. Again, such a reaction does not normally occur in nature, so the hydrogen system (18) is considered to be the most reducing of the natural redox systems.

Based on these considerations, it is possible to say that all redox reactions in paddy soils are those occurring in water and, have their standard redox potential between those of the oxygen system and the hydrogen system, at or near neutral reaction, pH 7. This is also justified because, as stated in the preceding chapter, soil (solution) pH after submergence tends to converge to near-neutrality. Therefore, in the following discussion, it is convenient to use the standard redox potential and pE° values recalculated for pH 7. The $E°_7$ and pE°$_7$ in Table 4.1 are the recalculated values, substituting 7 for pH in equations (10) and (16). The order of $E°$ values for the O_2-H_2O system and the NO_3^--N_2 system is reversed in the order of $E°_7$ values, and this more reasonably explains the denitrification reactions occurring in paddy fields. Thus, the following discussion assumes that all the reactions occur in neutral natural water, so that the relevant values of the redox potential and electron activity are those measured at pH 7, or $E°_7$ and pE°$_7$.

4.3.2 Redox reactions in paddy fields

Important redox systems that operate in paddy fields are listed in Table 4.1. A brief explanation for each of the systems follows.

1) Oxygen system

The oxygen system is the most oxidizing system in submerged paddy fields. In the presence of oxygen, plant roots and microbes obtain their energy by aerobic respiration, that is, by oxidizing carbohydrates with oxygen. Thus, the relevant redox systems are the oxygen system and the carbohydrate system, shown here as:

$$(1/4)O_2 + H^+ + e = (1/2) H_2O \qquad\qquad pE°_7 = 13.7$$
$$\text{and} \quad (1/4)CO_2 + H^+ + e = (1/4)CH_2O + (1/4)H_2O \qquad pE°_7 = -8.2$$

By combining the two half reactions the following complete reaction is obtained:

$$\begin{array}{lr}
 & \log K_7 \\
(1/4)\ O_2 + H^+ + e = (1/2)\ H_2O & 13.7 \\
(1/4)\ CH_2O + (1/4)\ H_2O = (1/4)\ CO_2 + H^+ + e & 8.2 \\
\hline
(1/4)\ CH_2O + (1/4)\ O_2 = (1/4)\ CO_2 + (1/4)H_2O & 21.9
\end{array}$$

The value of $\log K_7$, the log K at pH 7 or in natural water, was obtained from pE°$_7$ using the relationship, pE° = $(1/n)$ log K. This reaction yields 125 kJ of energy per 1mole of electron transfer at pH 7, as calculated below:

$$\Delta G°_7 = -RT \ln K_7$$
$$= -2.3\ RT \log K_7$$

$$= -5.7 \log K_7 \, (\text{kJ mol}^{-1})$$
$$= -125 \text{ kJ mol}^{-1} \, (= -29.8 \text{ kcal mol}^{-1})$$

As is evident from the method of calculation, this is the most energy efficient of all the reactions that utilize carbohydrates as the substrate. Therefore, oxygen is first used up when the soil is submerged.

In a soil solution that is in equilibrium with the atmosphere (that is, the partial pressure of oxygen, or P_{O2} equals 0.21 atm and the partial pressure of carbon dioxide, or P_{CO2} equals 0.0003 atm), the solution pH becomes 5.7 (at 25°C)[3] and the Eh is calculated to be 0.88 V, using equation (10). A change in the partial pressure of oxygen affects Eh very little, for example, only 15 mV for a ten times difference. Therefore, for an oxidative system of well-drained upland soil, the Eh cannot be a good measure of the redox condition. Furthermore, oxygen dissolved in water does not allow an exact measurement of electrode potential because of the very slow rate of electron exchange at the electrode surface, or what is called 'slow electrode'. An Eh of soil measured before submergence normally gives a potential of around 600 mV, which does not reflect the exact Eh of the oxygen system, 0.82V at pH 7.

2) Nitrogen system

The system consisting of NO_3^- and N_2 comes next in the descending order of $E°_7$ or $pE°_7$ (as in Table 4.1). The following equation is used to express the transfer of a single electron in the system:

$$(1/5) \, NO_3^- + (6/5) \, H^+ + e = (1/10) \, N_2 + (3/5) \, H_2O \qquad pE°_7 = \log K_7 = 12.6$$

Denitrification is a combination of this with the carbohydrate system, resulting in:

$$(1/4) \, CH_2O + (1/5) \, NO_3^- + (1/5) \, H^+ = (1/4) \, CO_2 + (1/10) \, N_2 + (7/20) \, H_2O$$
$$\log K_7 = 20.8$$

This reaction is often called 'nitrate respiration'. Energy yielded by the denitrification reaction is $\Delta G°_7 = -119$ kJ. The difference in the energy yield from the oxygen system is so small that denitrification is mediated by facultative anaerobes called 'denitrifiers' even at an early stage of soil submergence.

It is possible to consider a reduction from NO_3^- to NH_4^+, which is a reaction occurring in plants with the help of an enzyme, nitrate reductase. Is it then possible to have the following reaction in reduced soil?

$$(1/8)\ NO_3^- + (5/4)\ H^+ + e = (1/8)\ NH_4^+ + (3/8)\ H_2O \qquad pE^\circ_7 = 6.2$$

In fact, it is known that only a small percentage of nitrate may be reduced to NH_4^+ in reduced soil. This is most likely because of a competing reduction reaction that leads to denitrification, which has higher energy efficiency, as evident from the higher pE°_7 value. Yamane (1957) showed that NO_3^- added to reduced soil is almost quantitatively reduced to N_2, but not to NH_4^+. However, in an experiment with ^{15}N labeled 'NO_3-N' it was shown that, if amended with glucose or pre-incubated with glucose, 9 to 36% of N was converted to NH_4-N after four days of incubation in the Ar (argon) atmosphere (Buresh and Patrick, 1978). As a matter of fact, many microbial species are known to mediate a dissimilatory reduction of NO_3^- to NH_4^+ (Knowles, 1982).

The above reaction NO_3^-/NH_4^+ is reversed in the presence of oxygen and NH_4-N is oxidized to NO_3-N. This is the nitrification reaction mediated by chemoautotrophs that utilize the energy yielded by the following reaction, for which ΔG°_7 is -43 kJ. Nitrification occurs in a thin oxidized layer at the soil/water interface that is formed a few to several weeks after the paddy soil is submerged (see Chapter 3).

$$\log K_7$$

$$(1/4)\ O_2 + H^+ + e = (1/2)\ H_2O \qquad\qquad 13.7$$

$$\underline{(1/8)\ NH_4^+ + (3/8)\ H_2O = (1/8)\ NO_3^- + (5/4)\ H^+ + e \qquad -6.2}$$

$$(1/8)\ NH_4^+ + (1/4)\ O_2 = (1/8)\ NO_3^- + (1/4)\ H^+ + (1/8)\ H_2O \qquad 7.5$$

3) Manganese system

Manganese occurs in nature as oxide minerals with valence states of $+2$, $+3$, and $+4$. This complexity makes theoretical treatments of Mn systems difficult, because those taking part in the redox equilibria in the soil are mixtures of these different compounds. If MnO_2 is taken as a representative of manganese oxides, the manganese system is written as:

$$(1/2)\ MnO_2(s) + 2H^+ + e = Mn^{2+} + H_2O \qquad pE^\circ_7 = 6.8$$

Ponnamperuma et al. (1969) found that in soils with a high content of Mn, the standard redox potential E° was 0.90 V, instead of the theoretically derived 1.23 V. The same authors showed that in a system with a high partial pressure of CO_2, manganous carbonate $MnCO_3$ is precipitated under anaerobic conditions from Mn_3O_4 (hausmannite), which is also stable in an anaerobic medium with a lower CO_2 partial pressure:

$$(1/2)\ Mn_3O_4(s) + (3/2)\ CO_2 + H^+ + e = (3/2)\ MnCO_3(s) + (1/2)\ H_2O$$

The experimentally derived value of $E°$ for the above reaction was 0.64 V, whereas the theoretical value was 1.10 V. The discrepancy reveals a lower reactivity of Mn_3O_4 formed in the soil compared with a pure hausmannite mineral.

Stumm and Morgan (1970) proposed the following redox half reaction for the manganese system in carbonated natural water:

$$(1/2)\ MnO_2(s) + (1/2)\ HCO_3^-\ (at\ 10^{-3}\ M) + (3/2)\ H^+ + e = (1/2)\ MnCO_3(s) + H_2O$$

The $pE°_7$ value for the above reaction is 8.5 and this, in combination with the carbohydrate system, gives a log K_7 of 16.7 and a $\Delta G°_7$ of -95.2 kJ mol^{-1}. The partial pressure of CO_2 assumed in this reaction is a few tens of times its atmospheric partial pressure, and is a reasonable assumption even for the reduced soil system.

4) Iron system

Iron is by far the most important constituent that undergoes a redox reaction in the soil. The free iron oxide content of soil usually exceeds that of free manganese oxide by a factor larger than 10, so the ferric/ferrous pair is almost the sole electroactive redox system in submerged paddy soils. The most important iron system is often depicted as:

$$Fe(OH)_3(s) + 3H^+ + e = Fe^{2+} + 3H_2O \qquad pE°_7 = -3.1$$

As $E°$ is 1.06 volt and $pE°$ is 17.9, the Eh and pE of the reaction may be expressed as:

$$Eh = 1.06 - 0.059\ \log(Fe^{2+}) - 0.177\ pH$$
$$pE = 17.9 - \log(Fe^{2+}) - 3\ pH$$

Stumm and Morgan (1970) prefer to write the iron system in natural water as:

$$FeOOH(s) + HCO_3^-\ (at\ 10^{-3}\ M) + 2H^+ + e = FeCO_3(s) + 2H_2O \qquad pE°_7 = -1.67$$

FeOOH is either goethite (prevalent in both oxidized and reduced soils as iron oxide mottles), or lepidocrocite (occurring only in soils that tend to have frequent water saturation). As stated in the preceding section, the presence of bicarbonate at $10^{-3}\ M$ is a reasonable assumption. Thus, it also seems reasonable to consider the FeOOH/FeCO$_3$ redox pair for a reduced soil system. As stated below, the occurrence of FeCO$_3$(s) (or siderite) in some paddy soils is also confirmed (Yamazaki and Yoshizawa, 1961).

The above reaction in combination with the carbohydrate oxidation yields a log K_7 of 6.53 and a $\Delta G°_7$ of −37.2 kJ per mole of electron transfer.

For the FeOOH/FeCO$_3$ system in natural water, $E°$ and pE° are 0.73 V and 12.33, respectively, as given by Stumm and Morgan (1970). Therefore, the *Eh* and pE may be written as follows:

$$Eh = 0.73 + 0.059 \log (HCO_3^-) - 0.118 \text{ pH}$$
$$pE = 12.33 + \log (HCO_3^-) - 2 \text{ pH}$$

At pH 7 and (HCO$_3^-$) concentration of 10^{-3} *M*, the FeOOH/FeCO$_3$ system attains redox equilibrium at an *Eh* of about −0.27 V or a pE of −4.67.

Ponnamperuma (1972) considers the solid ferrous compound that determines Fe^{2+} solubility to be ferroso-ferric or ferrosic hydroxide Fe$_3$(OH)$_8$ rather than FeCO$_3$, because the latter precipitates only very slowly. In contrast, Lindsay (1979) considers that ferrosic hydroxide, as hypothesized by Ponnamperuma, has too high a solubility to be the solid phase in a strongly reduced condition with a high CO$_2$ partial pressure. A more detailed argument on this is presented in Chapter 6 concerning carbonate equilibrium.

5) Fermentation of organic substrates

In an anaerobic medium, microorganisms utilize organic matter as the electron acceptor, thereby simultaneously oxidizing part of the same organic matter. Such a biochemical reaction is called fermentation. In methanol fermentation, carbohydrate is reduced to methanol and another part of the carbohydrate is oxidized to formic acid or even to carbon dioxide.

	log K_7
(1/2) CH$_2$O + H$^+$ + e = (1/2) CH$_3$OH	−3.01
(1/2) CH$_2$O + (1/2) H$_2$O = (1/2) HCOO$^-$ + (3/2) H$^+$ + e	7.68
CH$_2$O + (1/2) H$_2$O = (1/2) HCOO$^-$ + (1/2)H$^+$ + (1/2)CH$_3$OH	4.67

The overall reaction produces a $\Delta G°_7$ of −26.6 kJ mol^{-1}, which is slightly less than the energy yielded by a reaction to form methanol and CO$_2$:

$$(3/4) \text{ CH}_2O + (1/4) \text{ H}_2O = (1/4) \text{ CO}_2 + (1/2) \text{ CH}_3OH, \qquad \log K_7 = 5.2$$

for which $\Delta G°_7 = -40.5$ kJ mol^{-1}.

6) Sulfur system

The most common mineral sulfur species in aerobic soils is sulfate, SO_4^{2-}. This will be reduced to sulfide in a strongly reductive condition by strict anaerobes, that is, sulfate reducers belonging to genera such as *Desulfovibrio* and *Desulfotomaculum*. In natural water at pH 7 the following equation holds:

$$(1/8)SO_4^{2-} + (9/8)H^+ + e = (1/8)HS^- + (1/2) H_2O \qquad pE^\circ_7 = -3.75$$

As is evident from the pE°_7 value, this reaction in combination with the carbohydrate oxidation yields a log K_7 of 4.45 and only a small amount of energy ($\Delta G^\circ_7 = -25.4$ kJ mol^{-1}). Sulfate reducers live on this energy, and this energy yielding reaction is sometimes called the 'sulfate respiration'.

7) Methane fermentation

In an extremely reductive condition, methane evolves along two pathways. One is a decarboxylation from acetic acid, which is abundant in reduced soils as a major product of fermentation, and the other is a reduction of carbon dioxide, which may be expressed as:

$$(1/8) CO_2(g) + H^+ + e = (1/8) CH_4(g) + (1/4) H_2O \qquad pE^\circ_7 = -4.14$$

As evident from the more negative value of pE°_7, the energy yielded by this reaction in conjunction with the carbohydrate oxidation:

$$(1/4) CH_2O = (1/8) CH_4 + (1/8) CO_2 \qquad \log K_7 = 4.06$$

is even lower than that from the sulfate reduction ($\Delta G^\circ_7 = -23.1$ kJ mol^{-1}).

The decarboxylation of acetic acid may be written as $CH_3COOH = CH_4 + CO_2$. This reaction may be considered to consist of the following two reactions (Stumm and Morgan, 1981):

$$CH_3COOH + 2H_2O = 2CO_2 + 8H^+ + 8e$$
$$CO_2 + 8H^+ + 8e = CH_4 + 2H_2O$$

8) Hydrogen system

There are laboratory findings that prove the existence of a transient process of hydrogen gas evolution from submerged soils (Asami, 1969; Yamane and Sato, 1968). The process occurs in soil with large amounts of readily decomposable organic matter during the initial stage of soil reduction, and produces a deep notch in the Eh-time curve at its starting point. As stated earlier, the hydrogen system is the most strongly reductive of all the natural aquatic systems:

$$H^+ + e = (1/2) H_2(g)$$

The reaction is taken as the standard for measuring the electric potential of other redox systems, so has a $E°$ of 0 V. At a pH of 7, or in natural water, $E°_7$ is −0.41 V and $pE°_7$ is −7. The overall reaction of hydrogen evolution through carbohydrate decomposition:

$$(1/4) CH_2O + (1/4) H_2O = (1/4) CO_2 + (1/2) H_2 \qquad \log K_7 = 1.2$$

yields very little energy, $\Delta G°_7 = -6.8$ kJ mol^{-1}.

4.4 Eh-pH diagram and pE-pH diagram

An Eh-pH diagram or a pE-pH diagram is also called an 'Eh (or pE)-pH stability field diagram'. It shows in a comprehensive way how protons and electrons simultaneously shift equilibria between various chemical species of a reaction system. Also it can indicate which chemical species predominates under any given condition of Eh and pH.

The Fe^{III}- Fe^{II} system is the most important redox system in submerged soils. For the time being, only hydrated oxides and simple ions are considered for drawing an Eh-pH diagram (see Figure 4.2).

First, it is necessary to enumerate all the possible iron species and establish a stoichiometric relationship between any two of them.

(1) Fe^{3+} + e = Fe^{2+} $E° = 0.77$ V
       $Eh = 0.77 + 0.059 \log (Fe^{3+})/(Fe^{2+})$

(2) Fe^{3+} cannot exist in a pH range higher than about 3 and hydrolyzes spontaneously to precipitate ferric hydroxide:

Fe(OH)$_3$(s) = Fe^{3+} + 3OH$^-$ $\log K_{sp} = -37.2$
       pH = (1/3) {4.8 − log (Fe^{3+})}

Assuming the total activity of ionic species of iron (Fe^{3+} + Fe^{2+}) to be 10^{-3}, the pH at which Fe^{3+} ceases to exist in the solution is calculated to be 2.6.

(3) By combining the above two equations, the $Fe(OH)_3(s)$-Fe^{2+} equilibrium is depicted by:

$$Fe(OH)_3(s) + 3H^+ + e = Fe^{2+} + 3H_2O \qquad\qquad E^\circ = 1.06\ V$$
$$Eh = 1.06 - 0.059\ \log\ (Fe^{2+}) - 0.177pH$$

(4) Equilibrium between solid phase ferric hydroxide and ferrosic hydroxide is shown as:

$$3Fe(OH)_3(s) + H^+ + e = Fe_3(OH)_8(s) + H_2O \qquad\qquad E^\circ = 0.43\ V$$
$$Eh = 0.43 - 0.059\ pH$$

(5) Equilibrium between Fe^{2+} and solid ferrosic hydroxide is given by:

$$(1/2)Fe_3(OH)_8(s) + 4H^+ + e = (3/2)Fe^{2+} + 4H_2O \qquad\qquad E^\circ = 1.38\ V$$
$$Eh = 1.38 - 0.089\ \log\ (Fe^{2+}) - 0.236pH$$

(6) Equilibrium between ferrosic hydroxide and ferrous hydroxide:

$$(1/2)Fe_3(OH)_8(s) + H^+ + e = (3/2)Fe(OH)_2(s) + H_2O \qquad\qquad E^\circ = 0.20\ V$$
$$Eh = 0.20 - 0.059pH$$

(7) Ferrous ion solubility as controlled by ferrous hydroxide:

$$Fe(OH)_2(s) = Fe^{2+} + 2OH^- \qquad\qquad \log K_{sp} = -14.7$$
$$pH = (1/2)\{13.3 - \log\ (Fe^{2+})\}$$

Assuming again (Fe^{2+}) to be 10^{-3}, the pH is calculated to be 8.1. The Eh-pH relationships obtained for the preceding reactions are illustrated in Figure 4.2 for the activity of dissolved iron species being 10^{-3}. The periphery of the stability field is delimited with the relationships for the oxygen ($Eh = 1.23 + 0.015\ \log\ P_{O2} - 0.059pH$) and the hydrogen system ($Eh = 0 - 0.03\ \log\ P_{H2} - 0.059\ pH$), assuming the respective gas at 1 atm partial pressure.

However, it is more realistic to take the carbonate system into consideration, as the equilibria of iron species in reduced soils are always strongly controlled by the carbonate system (see Chapter 6). A pE-pH diagram for the Fe-CO_2-H_2O system is provided below, quoting relevant reactions and constants from Yamane (1970). In this system, $FeCO_3$ must be assumed to be a stable solid phase, as should be various iron hydroxides, for

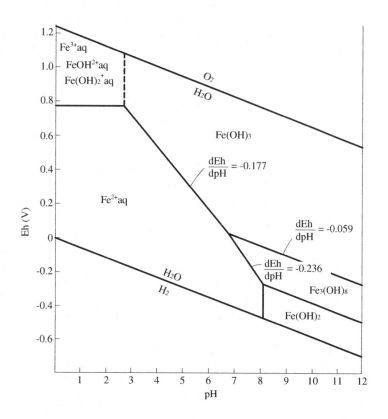

Figure 4.2
Eh-*pH diagram of Fe-H₂O system in submerged soil (25°C, total concentration
of iron species at 10⁻³ M) (Source: Ponnamperuma et al., 1967)*

example, $Fe(OH)_3$, $Fe_3(OH)_8$, and $Fe(OH)_2$ (see Chapter 6). Furthermore, the concentration of total carbonate species $(C_T = \{(H_2CO_3) + (HCO_3^-) + (CO_3^{2-})\})$ is assumed to be $10^{-3} M$, or 100 times the equilibrium concentration of pure carbonate solution in a normal atmosphere. The concentration of dissolved iron species is assumed to be $10^{-3} M$ as before. These assumptions would be more realistic for the natural paddy soil systems.

The equations needed are:

(1) $Fe^{3+} + e = Fe^{2+}$ $\qquad\qquad pE^\circ = 13.0$
$\quad pE = 13.0 + \log [(Fe^{3+})/(Fe^{2+})]$

(2) $Fe^{3+} + H_2O = FeOH^{2+} + H^+$ $\qquad\qquad \log K = -2.4$
$\quad pH = 2.4 - \log [(Fe^{3+}) / (FeOH^{2+})]$

(3) $FeOH^{2+} + H^+ + e = Fe^{2+} + H_2O$ $\qquad$ $pE^o = 15.5$

$\qquad pE = 15.5 - \log [(Fe^{2+}) / (FeOH^{2+})] - pH$

(4) $FeOH^{2+} + 2H_2O = Fe(OH)_3(s) + 2H^+$ $\qquad$ $\log K = -2.4$

$\qquad pH = 1.2 - (1/2)\log (FeOH^{2+})$

(5) $Fe(OH)_3(s) + 3H^+ + e = Fe^{2+} + 3H_2O$ $\qquad$ $pE^o = 17.6$

$\qquad pE = 17.6 - \log (Fe^{2+}) - 3pH$

(6) $3Fe(OH)_3(s) + H^+ + e = Fe_3(OH)_8(s) + H_2O$ $\qquad$ $pE^o = 7.3$

$\qquad pE = 7.3 - pH$

(7) $(1/2)Fe_3(OH)_8(s) + 4H^+ + e = (3/2)Fe^{2+} + 4H_2O$ $\qquad$ $pE^o = 23.2$

$\qquad pE = 23.2 - (3/2)\log (Fe^{2+}) - 4pH$

(8) $(1/2)Fe_3(OH)_8(s) + H^+ + e = (3/2)Fe(OH)_2(s) + H_2O$ $\qquad$ $pE^o = 3.3$

$\qquad pE = 3.3 - pH$

(9) $Fe^{2+} + 2H_2O = Fe(OH)_2(s) + 2H^+$ $\qquad$ $\log K = -13.3$

$\qquad pH = 6.6 - (1/2)\log Fe^{2+}$

(10-1) $Fe^{2+} + H_2CO_3 = FeCO_3(s) + 2H^+$ $\qquad$ $\log K = -6.0$

$\qquad pH = 3.0 - (1/2)\log (Fe^{2+}) - (1/2)\log (H_2CO_3)$

(10-2) $Fe^{2+} + HCO_3^- = FeCO_3(s) + H^+$ $\qquad$ $\log K = 0.3$

$\qquad pH = -0.3 - \log (Fe^{2+}) - \log (HCO_3^-)$

(11-1) $Fe(OH)_3(s) + H_2CO_3 + H^+ + e = FeCO_3(s) + 3H_2O$ $\qquad$ $pE^o = 11.8$

$\qquad pE = 11.8 + \log (H_2CO_3) - pH$

(11-2) $Fe(OH)_3(s) + 2H^+ + HCO_3^- + e = FeCO_3(s) + 3H_2O$ $\qquad$ $pE^o = 18.2$

$\qquad pE = 18.2 + \log (HCO_3^-) - 2pH$

(12-1) $(1/2) Fe_3(OH)_8(s) + (3/2) HCO_3^- + (5/2)H^+ + e = (3/2)FeCO_3(s) + 4H_2O$

$\qquad\qquad\qquad\qquad\qquad\qquad\qquad\qquad\qquad pE^o = 23.6$

$\qquad pE = 23.6 + (3/2)\log (HCO_3^-) - (5/2)pH$

(12-2) $(1/2)Fe_3(OH)_8(s) + (3/2)(CO_3^{2-}) + 4H^+ + e = (3/2)FeCO_3(s) + 4H_2O$

$\qquad\qquad\qquad\qquad\qquad\qquad\qquad\qquad\qquad pE^o = 39.1$

$\qquad pE = 39.1 + (3/2)\log (CO_3^{2-}) - 4pH$

(13-1) $Fe(OH)_2(s) + H^+ + HCO_3^- = FeCO_3(s) + 2H_2O$ $\log K = 13.6$
$$pH = 13.6 + \log (HCO_3^-)$$

(13-2) $Fe(OH)_2(s) + CO_3^{2-} + 2H^+ = FeCO_3(s) + 2H_2O$ $\log K = 23.8$
$$pH = 11.9 + (1/2)\log (CO_3^{2-})$$

$E°$ values in Yamane's paper (1970) were converted to pE° to draw the pE-pH diagram. It is assumed here that the carbonate species changes from H_2CO_3 to HCO_3^- at pH 6.4, and from HCO_3^- to CO_3^{2-} at pH 10.3, and that only one carbonate species exists in each pH range.

Figure 4.3 shows the relationships among the components of the Fe-CO_2-H_2O system considered above. The shaded area is the stability field of $FeCO_3(s)$. The broken lines within the shaded area are the boundaries when carbonate is not taken into consideration. In this diagram, partially

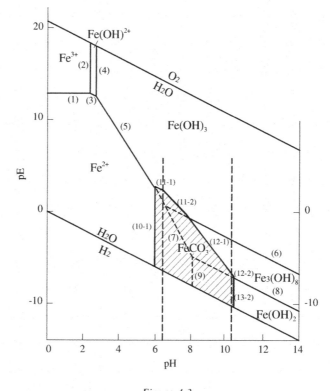

Figure 4.3
pE-pH diagram of Fe-CO$_2$-H$_2$O system in submerged soil (25°C, total concentration of both carbonate and iron species at 10^{-3} M) (Source: Yamane, 1970, partly modified)

hydrolyzed ferric iron species, $FeOH^{2+}$, is considered, but from equations (2) and (4) it is known that when the dissolved iron species have concentrations higher than $10^{-2.4}$ M, the field for $FeOH^{2+}$ disappears from the diagram. The equations (10-2) and (13-1) are not expressed in the figure because these equations are not relevant to the respective pH range with respect to the carbonate species. If the total iron and carbonate concentrations are lower than assumed here, they would be used instead of (10-1) and (13-2).

From Figure 4.3 it is confirmed that ferrous carbonate or siderite ($FeCO_3$), is a more probable form of the solid phase in the ordinary Eh (or pE)-pH ranges of paddy soils than ferrosic hydroxide [$Fe_3(OH)_8$] under a high carbonate condition as found in paddy soils.

Notes

(1) Liquid junction potentials arise from the transfer of ionic species through the transition region that divides the test solution from the reference electrode solution. The liquid junction potential makes a contribution to the electromotive force of the cell; it increases with increasing difference between the two solutions that form a single junction. The liquid junction potential can often be kept small by using a concentrated salt bridge (quoted from Stumm and Morgan, 1981).

(2) In recent years, the Ag-AgCl electrode has been used commonly as the reference electrode, instead of the calomel electrode. The potential values (in V) of the Ag-AgCl electrode are:

Temperature °C	0.1 M KCl	3.3 M KCl	saturated KCl
10	–	0.217	0.214
15	–	0.214	0.209
20	–	0.210	0.204
25	0.290	0.206	0.199

(3) The pH of water in equilibrium with an atmospheric carbon dioxide concentration of 0.03% by volume may be calculated as follows:
The partial pressure of CO_2 in the atmosphere, P_{CO2}, is

$$P_{CO2} = 0.0003 \text{atm or } 10^{-3.5} \text{atm.}$$

The dissolved carbonate concentration in water that is in equilibrium with atmospheric CO_2 is:

$$(H_2CO_3) = K_H \cdot P_{CO2} = 10^{-5}(M) \qquad \text{(i)}$$

where K_H is the Henry's law constant, the value for CO_2 being $10^{-1.5}$ at 25°C.

The first and second dissociation constants of H_2CO_3 are:

$$K_1 = (H^+)(HCO_3^-) / (H_2CO_3) = 10^{-6.4} \qquad \text{(ii)}$$
$$K_2 = (H^+)(CO_3^{2-}) / (HCO_3^-) = 10^{-10.3} \qquad \text{(iii)}$$

The charge balance of the pure H_2CO_3 solution is expressed as:

$$(H^+) = (HCO_3^-) + 2(CO_3^{2-}) + (OH^-)$$

In an acidic condition:

$$(H^+) \doteqdot (HCO_3^-) \qquad \text{(iv)}$$

Therefore, from equations (ii) and (iv) we obtain:

$$(H^+)^2 / (H_2CO_3) = 10^{-6.4} \qquad \text{(v)}$$

(H^+) may be calculated from (v) by substituting 10^{-5} in (i) for (H_2CO_3):

$$(H^+) = 10^{-5.7} (M)$$

Thus, the pH value of natural water in equilibrium with atmospheric CO_2 is approximately 5.7. This value is also used to define acid rain.

References

Asami, T. 1969. Initial rapid drop of oxidation-reduction potentials in submerged paddy soils. *J. Sci. Soil & Manure, Japan*, 40: 429–430. (In Japanese).

Buresh, R.J. and Patrick, W.H. 1978. Nitrate to ammonium in anaerobic soil. *Soil Sci. Soc. Amer. J.*, 42: 913–918.

Knowles, R. 1982. Denitrification. *Microbiol. Rev.*, 46: 43–70.

Lindsay, W.L. 1979. *Chemical Equilibria in Soils*. John Wiley & Sons, N.Y.

Ponnamperuma, F.N. 1972. The chemistry of submerged soils. *Adv. in Agron.* 24: 29–96.

Ponnamperuma, F.N., Loy, T.A. and Tianco, E.M. 1969. Redox equilibrium in flooded soils. II, The manganese oxide system, *Soil Sci.*, 108: 48–57.

Sillen, L.G. 1967. Master variables and activity scales. In Equilibrium Concepts in Natural Water Systems, *Advances in Chemistry Series*, No. 67, pp. 45–69. American Chemical Society, Washington, D.C. (Cited from Stumm and Morgan, 1981).

Stumm, W. and Morgan, J.J. 1970. *Aquatic Chemistry: An Introduction Emphasizing Chemical Equilibria in Natural Waters.* John Wiley & Sons, N.Y.

Stumm, W. and Morgan, J.J. 1981. *Aquatic Chemistry: An Introduction Emphasizing Chemical Equilibria in Natural Waters. 2nd Ed.* John Wiley & Sons, N.Y.

Yamane, I. 1957. Nitrate reduction and denitrification in flooded soils. *Soil & Plant Food*, 3: 100–103.

Yamane, I. 1961. Manometric micro-analysis of gas mixtures and some problems of gas formation in rice paddy soil. *Bull. Inst. Agric. Res. Tohoku Univ.*, 12: 262–363. (In Japanese).

Yamane, I. 1970. *Eh*-pH diagrams of iron systems in relation to flooded soils. *Rep. Inst. Agric. Res. Tohoku Univ.*, 21: 39–63.

Yamane, I. and Sato, K. 1968. Initial rapid drop of oxidation-reduction potential in submerged air-dried soils. *Soil Sci. Plant Nutr.*, 14: 68–72.

Yamane, I. and Sato, K. 1968. On the measurement of redox potentials of submerged soils. *J. Sci. Soil & Manure, Japan*, 39: 535–542. (In Japanese).

Yamazaki, T. and Yoshizawa, T. 1961. Mottles and concretions of ferrous carbonate ($FeCO_3$) occurring in paddy soil profiles, Part 1. *Bull. Hokuriku Agric. Expt. Sta.*, 2: 1–16. (In Japanese).

Zobell, C.E. 1946. *Bull. Amer. Ass. Petrol. Geol.*, 30: 477–513. (Cited from Ponnamperuma, 1972).

Chapter 5: Fundamental Biological and Biochemical Reactions in Submerged Paddy Soils

5.1 Introduction

Microorganisms mediate almost all of the chemical reactions that were referred to in the preceding chapter. Therefore, in order to understand the processes taking place in paddy soils it is imperative to understand the microbiological activities in the soil.

It is pertinent here to recall the features of the microbial population in paddy soils. Ishizawa and Toyoda (1964) summarized the results of their studies on microbial populations in Japanese arable lands (Table 5.1). The results for paddy soils used measurements taken during the non-submerged period. Compared with upland soils, paddy soils have more aerobes (both obligate and facultative), and less actinomycetes and fungi in the surface, subsurface and subsoil. Also, sulfate reducers (obligate anaerobes) and denitrifiers (facultative anaerobes) are present in much greater numbers in paddy soils than in upland soils. This suggests that an anaerobic environment is maintained in paddy soils to some extent, even after draining of the surface water.

Table 5.1 Distribution of microorganisms in paddy and upland soils ($\times\,10^4\,g^{-1}$)

Flora	Paddy (mean of 18 samples)			Upland (mean of 26 samples)		
	1st layer	2nd layer	3rd layer	1st layer	2nd layer	3rd layer
Aerobes	3000	1310	837	2185	628	164
Actinomycetes	200	88	38	477	172	35
Fungi	8.5	1.6	0.6	23.1	4.3	1.1
Anaerobes	232	112	22	147	57	16
SO_4-reducers	7.9	1.6	0.4	0.1	0.06	3
Denitrifiers	29.7	16.4	12.2	4.7	2.7	–
Nitrifiers	1.1	–	–	7.0	5.3	0.05

(Source: Ishizawa and Toyoda, 1964)

Changes in microbial populations during the submerged period were discussed briefly in Chapter 3. Submergence alters the character of microbial flora in the soil. Aerobes reach their peak numbers a few days after the soil is submerged, but they are then replaced by facultative anaerobes, which in turn are superseded by anaerobes (Takai, 1969). These microorganisms play a very important role in the redox transformation of various substances in paddy soils, both during submergence and after drainage.

5.2 Various metabolic pathways operating in paddy soils

As stated earlier, a few to several weeks after submergence, differentiation of a thin oxidized surface layer occurs, and the surface water and the thin surface layer maintain an aerobic condition throughout the rest of the submerged period due to the continuous, though very slow, diffusion of the air through the water. The oxygen that reaches the soil/water interface is consumed rapidly by aerobes or by reducing substances, such as ferrous and manganous ions, which diffuse upward from the underlying reduced layer.

5.2.1 Aerobic metabolisms

The concentration of oxygen that limits respiration of aerobic bacteria is reported to be 1% of the air-saturated level, under standard temperature and pressure conditions. Therefore, until the oxygen concentration reaches this level, aerobic metabolisms predominate even in submerged soils.

Common aerobic heterotrophic bacteria in the soil decompose organic matter to its final decomposition products, CO_2, H_2O, and NH_4^+, by aerobic respiration using molecular oxygen as the terminal electron acceptor. The reaction:

$$C_6H_{12}O_6 + 6O_2 = 6CO_2 + 6H_2O \qquad \Delta G° = -688 \text{ kcal mol}^{-1}$$

is the reverse of the photosynthesis, and the energy released from the reaction is utilized by heterotrophic microorganisms. (It has been noted already in Chapter 4 that energy released by a transfer of one mole of electron in aerobic respiration was calculated to be 29.8 kcal. In the above formula, twenty-four moles of electrons are involved in the reaction and the calculated energy yield is 715 kcal mol^{-1}. There is a small discrepancy between these two numbers.)

There are many kinds of aerobic autotrophic bacteria which use molecular oxygen as an electron acceptor in the process of oxidation of inorganic compounds, thereby deriving their energy for assimilating carbon dioxide. They are called chemoautotrophic bacteria. The following list provides examples (Yoshida, 1978):

Nitrosomonas spp.[1]	$NH_4^+ + (3/2) O_2 = NO_2^- + 2H^+ + H_2O$
Nitrobacter spp.	$NO_2^- + (1/2) O_2 = NO_3^-$
Mn-oxidizing bacteria	$Mn^{2+} + O_2 = MnO_2$
Iron bacteria	$Fe^{2+} + H^+ + (1/4) O_2 = Fe^{3+} + (1/2) H_2O$
Thiobacillus spp.	$S + (3/2) O_2 + H_2O = H_2SO_4$
Beggiatoa spp.	$H_2S + (1/2) O_2 = S + H_2O$
Methane-oxidizing bacteria	$CH_4 + 2O_2 = CO_2 + 2H_2O$
Hydrogenomonas spp.	$H_2 + (1/2) O_2 = H_2O$

Nitrification is a very important process in the aerobic transformation of NH_4^+ to NO_3^- mediated by chemoautotrophs both in upland soils and paddy soils (see Chapter 4). The importance of bacteria in manganese oxidation has been documented for Japanese paddy soils (Motomura, 1966; Kamura and Nishitani, 1977). The occurrence of a reddish brown film of hydrated iron oxides in the surface water of permanently water-saturated paddy soils (wet paddy field soils) is attributed to the activity of iron bacteria (*Ferrobacillus* spp.). In a strongly acidic condition, say pH < 4.5, chemical oxidation of ferrous iron is very slow. However, *Thiobacillus ferrooxidans* oxidizes ferrous iron only in a very strongly acidic condition, between pH 2.4 and 3.5 (Yoshida, 1978). This should play an important role in the process of pyrite oxidation in acid sulfate soils. *Beggiatoa*, a filamentous bacterium, is known to colonize rice rhizosphere and oxidize hydrogen sulfide to elemental sulfur (Joshi and Hollis, 1977).

Methane and hydrogen gases are emitted by strongly reduced paddy soils. They are oxidized by methane oxidizing bacteria and *Hydrogenomonas* spp., respectively. These autotrophic bacteria play important roles, mainly in the aerobic surface water and in a thin oxidized layer at the soil/water interface. Anaerobic metabolisms predominate in the rest of the surface soil.

5.2.2 Anaerobic metabolisms

When the trapped and dissolved oxygen remaining after the submergence of paddy soils is completely used by aerobic microorganisms, the medium turns anaerobic and the aerobes become quiescent or die. Anaerobic microorganisms predominate in the medium and they live by anaerobic respiration, which is a redox process conducted first by facultative anaerobes and later by obligate anaerobes, in which an energy-yielding oxidation reaction is coupled with a reduction of inorganic oxides as an electron acceptor. These facultative and obligate anaerobes mediate the following reduction reactions (Yoshida, 1978) in accordance with the equations given in the preceding chapter:

Denitrification (denitrifying bacteria) $NO_3^- \rightarrow N_2$
Manganese reduction $MnO_2 \rightarrow Mn^{2+}$
Iron reduction $Fe_2O_3 \rightarrow Fe^{2+}$
Sulfate reduction (*Desulfovibrio* spp.) $SO_4^{2-} \rightarrow HS^-$
Methane fermentation (*Methanomonas* spp.) $CO_2 \rightarrow CH_4$

Indirect chemical reduction, by reducing metabolites originating from microbial processes, is considered more important in the reduction of manganic oxides than direct microbial reduction. Both microbial and chemical, or direct and indirect, reduction processes operate in the case of ferric oxides.

In these transformations, the oxidation of organic substrates is coupled with a reduction of mineral oxides; the standard redox potentials of which are given in Table 4.1. The reactions with higher standard potentials yield more energy than those with lower potentials, when coupled with the same oxidation reaction of organic matter. Thus, denitrifiers act at higher redox potential levels and gain energy more efficiently, while *Desulfovibrio* spp. and *Methanomonas* spp. act only at lower redox potential levels and gain much less energy (see Chapter 4).

The microbiological nature of these reduction processes has been verified by many studies. For example, the iron-reducing activity of submerged soils is strongly inhibited by sterilization of the soils using steam prior to submergence, or by the addition of inhibitors of bacterial activity, such as antibiotics, to the soil.

5.2.3 Fermentation

When there is sufficient oxygen to act as an electron acceptor, organic matter is oxidized or decomposed completely to carbon dioxide and water (and ammoniacal nitrogen) by heterotrophic microorganisms. When the partial pressure of oxygen is lowered, under submerged conditions, organic substances are partially degraded (as shown in the following reaction), producing alcohol from sugar and emitting carbon dioxide:

$$C_6H_{12}O_6 = 2CO_2 + 2C_2H_5OH$$

The energy released from the reaction can be utilized by the facultative anaerobes that mediate the reaction. Fermentation is defined as a biochemical process in which anaerobic microorganisms utilize organic materials both as the electron donor and acceptor. Thus, the end product is constituted by both reduced and oxidized organic compounds, such as alcohol and carbon dioxide.

It is believed that in both aerobic and anaerobic conditions, carbohydrates are metabolized along the same pathway up to pyruvic acid:

$$C_6H_{12}O_6 + 2NAD^+ + 2ADP + 2Pi = 2CH_3COCOOH + 2NADH + 2H^+ + 2ATP$$

where NAD is nicotinamide adenine dinucleotide. Pyruvic acid may be further transformed via different pathways depending on the conditions of the medium. Yoshida (1975) summarized the carbohydrate fermentation pathways, as in Figure 5.1. Many kinds of low molecular weight organic compounds have been detected in paddy soils: acetone, acetaldehyde, formic, acetic, propionic, butyric, valeric, caproic, lactic, oxaloacetic, malonic, fumaric and succinic acids. However, the main products of the fermentation of carbohydrates are carbon dioxide; the lower fatty acids, particularly acetic acid; and methane. Some fermentation products, such as butyric acid, are toxic to rice plants even at low concentrations. The

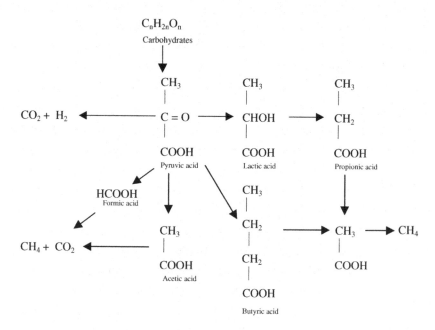

Figure 5.1
Diagram showing major organic acid and methane fermentation of carbohydrates in submerged soils (Source: Yoshida, 1975)

formation of organic acids and methane by fermentation is suppressed in the presence of nitrate, manganese oxides, or iron oxides in amounts large relative to the amount of decomposable organic matter (see below).

5.2.4 Soil enzymes

Most metabolic activities in paddy soils are mediated by enzymes that are secreted by microorganisms. Table 5.2 presents the major soil enzymes in three groups based on the element that they metabolise; carbon, nitrogen or phosphorus.

The pattern of seasonal fluctuations in soil enzyme activity is similar to that of aerobic bacteria. Factors that suppress enzyme activity are soil reduction during the rice-cropping season and low temperatures in winter (Kanazawa, 1982).

5.3 Biochemical processes in submerged paddy soils

In a model experiment using injection syringes, Takai et al. (1955) and Takai and Koyama (1956) studied the reduction process of soil in terms of its bacterial flora, Eh, gas composition, and mineral redox systems. (see Figures 5.2 and 5.3) The plate method bacterial count peaks, a few days after submergence; this represents the aerobes and some of the facultative anaerobes. Then, the deep layer method bacterial count peaks, representing facultative and obligate anaerobes. Finally, the number of sulfate reducing bacteria begins to increase towards the latter part of the experiment. Such changes in bacterial counts seem to correspond to the changes in redox potential, reduction of ferric compounds and sulfide formation.

In the first stage, the relative concentration of ferrous iron increases rapidly, as Eh_7 drops sharply. When Eh_7 is below -200 mV, active sulfide formation begins. Figure 5.3 shows the changes in gas components and mineral elements as a function of Eh_7. Oxygen and nitrate ions disappear, while Eh_7 remains between 400 and 200 mV. The reduction of ferric iron and the emission of CO_2 proceed actively as Eh_7 drops from 200 to -200 mV, while sulfide and methane formation only begins when Eh_7 is below -50 mV and -240 mV, respectively. Such a stepwise reduction process has been noted by many studies. Patrick and his coworkers studied the reduction process of many compounds using a special device that maintains the redox potential at a predetermined value (Eh-stat). They proposed the potentials at which the main oxidized components in submerged soils become unstable, as in Table 5.3 (Patrick and Reddy, 1978).

Table 5.2 Soil enzymes mediating carbon, nitrogen, and phosphorus metabolisms

Element	Function	Substrate	Enzyme	Biochemical reaction
Carbon	Hydrolysis	Starch	Amylase (α,β)	Hydrolysis of 1,4 glucosidic bonds of polyglucosans
		Cellulose	Cellulase	Hydrolysis of $\beta-1,4$-glucan links in cellulose
		Xylan	Xylanase	Hydrolysis of $\beta-1,4$-xylan links
		Inulin	Inulinase	Hydrolysis of $\beta-1,2$ fructan links
		Pectin	Pectinase	Pectin + $H_2O \rightarrow$ pectinic acid + galactose
		α-Glucosides	α-Glucosidase	α-R-glucoside + $H_2O \rightarrow$ R-OH + glucose
		β-Glucosides	β-Glucosidase	β-R-glucoside + $H_2O \rightarrow$ R-OH + glucose
		Sucrose	Saccharase	Sucrose + $H_2O \rightarrow$ glucose + fructose
		Neutral lipid	Lipase	Triglyceride + $3H_2O \rightarrow$ glycerol + fatty acid
	Oxidation	Glucose	Glucose oxidase	Glucose + $H_2O_2 \rightarrow$ gluconic acid
		Ascorbic acid	Ascorbic oxidase	Ascorbic acid + $H_2O_2 \rightarrow$ dehydro-ascorbic acid + H_2O
		Polyphenol	Polyphenol oxidase	$A + H_2O_2 \rightarrow$ oxidized $A + H_2O$
Nitrogen	Hydrolysis	Protein	Proteinase	Hydrolysis of protein to peptides and amino acids
		β-N-acetylgluco-saminides	β-N-acetylgluco-saminidase	β-R-N-acetylglucosaminide + H_2O $\rightarrow$ R-OH + N-acetylglucosamin
		Asparagine	Asparaginase	Asparagine + $H_2O \rightarrow$ asparagic acid + NH_3
		Glutamine	Glutaminase	Glutamine + $H_2O \rightarrow$ glutamic acid + NH_3
		Urea	Urease	$CO(NH_2)_2 + H2O \rightarrow 2NH_3 + CO_2$
	Denitrification	NO_3^-	Nitrate reductase	NO_3^- + reduced acceptor $\rightarrow NO_2^- + H_2O$ + acceptor
		NO_2^-	Nitrite reuctase	$2HNO_2$ + reduced acceptor $\rightarrow 2NO + 2 H_2O$ + acceptor
		NO	Nitric oxide reductase	NO + reduced acceptor $\rightarrow N_2$ + acceptor
Phosphorus	Hydrolysis	Phosphate-monoesterides	Phosphatase	Orthophosphoric monoester + $H_2O \rightarrow$ R-OH + PO_4^{3-}
		Inositol phosphates	Phytase	*myo*-Inositol hexaphosphate + $H_2O \rightarrow$ *myo*-inositol + $6H_3PO_4$
		Nucleotides	Nuclease	Dephosphorylation of nucleotides
		Metaphosphates	Metaphosphatase	Metaphosphoric acid + $H_2O \rightarrow 2H_3PO_4$
		Pyrophosphates	Pyrophosphatase	Pyrophosphoric acid + $H_2O \rightarrow 2H_3PO_4$

(Source: Kanazawa, 1982)

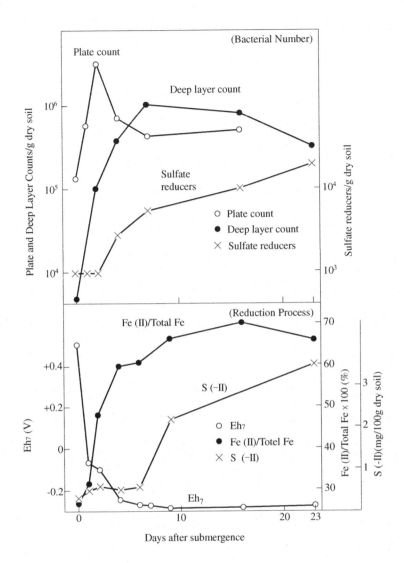

Figure 5.2
Reduction process and changes in bacterial numbers in submerged soils in a
syringe incubated at 35°C (Source: Takai et al.*, 1955)*

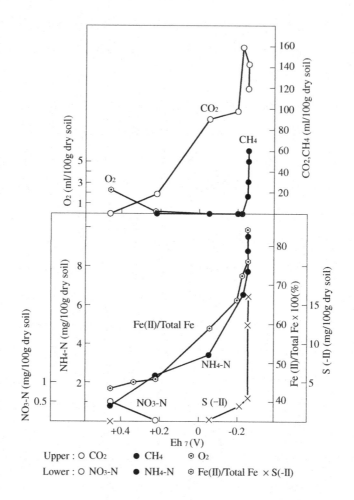

Figure 5.3

Relationship between Eh_7 *and material changes in submerged soil in a syringe incubated at 35°C for three weeks (Source: Takai and Koyama, 1956)*

Organic matter fermentation is suppressed by the addition of nitrate. Takai *et al.* (1957 a,b) studied the effect of ferric oxide and organic matter on the formation of organic acids and methane under submerged conditions. The soils used were as follows:

Soil	Texture	Fe_2O_3 %	NH_4-N mg/100g dry soil*
Nagano	CL	1.10	17.4
Sanage	CoSL	0.09	7.8

* Mineralized after seven weeks of incubation at 30°C under submergence.

Table 5.3   *Approximate redox potentials at which the main oxidized components in submerged soils become unstable*

Reaction	Redox potential (mV)		
$O_2 \rightarrow H_2O$	+380	~	+320
$NO_3^- \rightarrow N_2$; $Mn^{4+} \rightarrow Mn^{2+}$	+280	~	+220
$Fe^{3+} \rightarrow Fe^{2+}$	+180	~	+150
$SO_4^{2-} \rightarrow S^{2-}$	−120	~	−180
$CO_2 \rightarrow CH_4$	−200	~	−280

Source: Patrick and Reddy, 1978

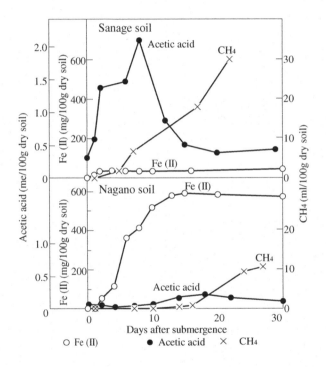

Figure 5.4
Changes in Fe^{II}, acetic acid and methane production in two submerged soils different in free iron and readily decomposable organic matter contents (Source: Takai et al., 1957)

The Nagano soil contains more ferric oxide and organic matter than the Sanage soil, which contains very little iron oxide relative to de-composable organic matter. The results of the experiment are shown in Figure 5.4. In the Sanage soil, iron reduction is completed by the second day of submergence. Acetic acid is formed vigorously from the very beginning of incubation, and methane is produced from the sixth day. It appears that active methane formation accompanies a rapid decline in acetic acid. The level of ferric iron reduction in the Nagano soil was seventeen times greater than that in the Sanage soil, and acetic acid and methane formation was apparently suppressed during the active reduction process of ferric iron.

Based on such experimental results, Takai *et al.* (1958) concluded that organic matter transformation coupled with ferric iron reduction does not produce fermentation products such as acetic acid, and this suppresses the production of methane that usually uses these fermentation products as a substrate (see Chapter 4). Kamura *et al.* (1963) further proved, in a ^{14}C tracer experiment, that during the decomposition of acetic acid that occurs in parallel with ferric iron reduction, methyl carbon is transformed to CO_2. Asami and Takai (1970) found that the production of volatile organic acids and methane is greatly suppressed when amorphous hydrated ferric oxide is added to submerged soils. During the first two weeks of submergence, they obtained a very high correlation coefficient between the release of CO_2 and the formation of Fe^{II}. This suggests that the oxidative breakdown of organic acids and the reduction of ferric oxides are linked. However, it has not been shown whether the reduction of Fe^{III} is accomplished by a direct enzymatic transfer of electrons to ferric iron, or by an indirect chemical reaction between bacterial metabolites and ferric iron.

Table 5.4 summarizes the biochemical reduction processes in submerged soils (Takai, 1978). Two major stages in reduction are differentiated: the aerobic and semi-anaerobic decomposition stage, and the anaerobic decomposition stage. The first stage consists of sequential reactions from oxygen removal to ferric iron reduction via nitrate reduction and manganese reduction. In the earlier part of the second stage, sulfate reduction occurs vigorously, followed by methane formation. During the first stage, Eh_7 drops from 500 mV to −200 mV, and ammonia and carbon dioxide are produced by the aerobic or semi-anaerobic decomposition of organic matter. At the end of the first stage, the Eh of the medium is sufficiently low for obligate anaerobes to become active, and organic substances are fermented to supply substrates to sulfate reducers and methanogenic bacteria.

It is important to note from the table that these phased reactions follow the sequence of decreasing efficiency of energy acquisition by micro-organisms, that is, from the redox reactions with high positive potentials to those with low or negative potentials.

Table 5.4 Reduction processes and microbial metabolisms in submerged soils

Stage of Reduction	Chemical transformation	Initial Eh_7 in soil, V	Expected pattern of energy metabolism	Formation of			Hypothetical pattern of organic matter decomp.	Redox state
				NH_4-N	CO_2	organic acid		
First	Disappearance of molecular oxygen	0.5	Aerobic respiration	Rapid process	Rapid process	Usually not accumulated without fresh organic matter	Aerobic and semi-anaerobic decomposition process	Oxidized
	Disappearance of nitrate	0.4	Nitrate reduction					Weakly reduced
	Formation of Mn^{II}	0.4	($Mn^{III,IV}$ reduction)					Moderately reduced
	Formation of Fe^{II}	0.2	(Fe^{III} reduction)					
Second	Formation of S^{-II}	0	Sulfate reduction	Slow Process	Slow to stagnant process	Early stage: rapid accumulation	Anaerobic decomposition process	Strongly reduced
	Formation of methane	−0.2	Methane fermentation			Advanced stage: rapid decrease		
	Formation of hydrogen	−0.2	Fermentation			Formation & decomposition of formic acid		

(Source: Takai, 1978, partly modified)

Notes

(1) Recent research has revealed that other genera like *Nitrosospira* and *Nitrosolobus* are more frequently identified in acid soils than *Nitrosomonas* (M. Hayatsu, personal communication).

References

Asami, T. and Takai, Y. 1970. Behavior of free iron oxide in paddy soils (Part 4). Relationship between reduction of free iron oxide and formation of gases in paddy soils. *J. Sci. Soil & Manure, Japan*, 41: 48–55. (In Japanese).

Ishizawa, S. and Toyoda, K. 1964. Microflora of Japanese soils. *Bull. Nat. Inst. Agric. Sci.*, B14: 203–284. (In Japanese).

Joshi, M.M. and Hollis, J.P. 1977. Interaction of Beggiatoa and rice plant: detoxification of hydrogen sulfide in the rice rhizosphere. *Sci.*, 195: 179–180. (Cited from Yoshida, 1978).

Kamura, T. and Nishitani, K. 1977. The effects of the soil reaction on the manganese oxidation (II). The oxidation mechanism of manganese in soils, Part 2. *J. Sci. Soil & Manure, Japan*, 48: 103–106. (In Japanese).

Kamura, T., Takai, Y. and Ishikawa, K. 1963. Microbial reduction mechanism of ferric iron in paddy soils (Part 1). *Soil Sci. Plant Nutr.*, 9: 171–175.

Kanazawa, S. 1982. Chapter 6 Biota in paddy soils. In Yamane, I. (Ed.) *Paddy Soil Science*, pp. 233–279. No-Bun-Kyo, Tokyo. (In Japanese).

Motomura, S. 1966. Studies on the oxidative sediments in paddy soils (Part 2). Estimation of the amount of manganese-oxidizing bacteria in paddy soils. *J. Sci. Soil & Manure, Japan*, 37: 263–268. (In Japanese).

Patrick, W.H. and Reddy, C.N. 1978. Chemical changes in rice soils. In IRRI (Ed.) *Soils and Rice*, pp.361–379. IRRI, Los Baños, Philippines.

Takai, Y. 1969. The mechanism of reduction in paddy soil. *Japan Agric. Res. Quarterly (JARQ)*, 4 (No.4): 20–23.

Takai, Y. 1978. Redox processes in the soil under submergence. In Kawaguchi, K. (Ed.) *Paddy Soil Science*, pp.23–55. Kodan-sha, Tokyo. (In Japanese).

Takai, Y. and Koyama, T. 1956. Microbial metabolism of paddy soils (Part 2). Composition of gases and organic acids contained in soils of paddy field. *J. Sci. Soil & Manure, Japan*, 26: 509–512. (In Japanese).

Takai, Y., Kamura, T. and Adachi, I. 1958. On dynamic behavior of iron compound in paddy soils. II. An improved method for determining Fe^{II}

in the waterlogged soil. *J. Sci. Soil & Manure, Japan*, 29: 216–220. (In Japanese).

Takai, Y., Koyama, T. and Kamura, T. 1955. Microbial metabolism of paddy soils, Part I. *J. Agric. Chem. Soc. Japan*, 29: 967–972. (In Japanese).

Takai, Y., Koyama, T. and Kamura, T. 1957 a. Microbial metabolism of paddy soils, Part III. Effect of iron and organic matter on the reduction process (1). *J. Agric. Chem. Soc. Japan*, 31: 211–215. (In Japanese).

Takai, Y., Koyama, T. and Kamura, T. 1957 b. Microbial metabolism of paddy soils, Part IV. Effect of iron and organic matter on the reduction process (2). *J. Agric. Chem. Soc. Japan*, 31: 215–220. (In Japanese).

Yoshida, T. 1975. Microbial metabolism of flooded soils. In Paul, E.A. and McLaren, A.D. (Eds.) *Soil Biochemistry*, vol. 3, pp. 83–121.

Yoshida, T. 1978. Microbial metabolism in rice soils. In IRRI (Ed.) *Soils and Rice*, pp.445–463. IRRI, Los Baños, Philippines.

Chapter 6: The Solubility and Redox Equilibria of Iron Systems in Submerged Paddy Soils

6.1 Introduction

As stated in the previous chapters, the following sequence of chemical and biochemical changes occurs in submerged soils:

- A rapid reduction of O_2, NO_3^-, Mn^{IV}, and Fe^{III} to H_2O, N_2, Mn^{II}, and Fe^{II}. This is accompanied by the emission of CO_2 and the formation of NH_4^+ and volatile fatty acids, especially acetic acid. *Eh* falls and pH rises during this stage of a rapid reduction.

- After a few to several weeks of submergence, the whole system achieves a more or less steady state; that is, the soil *Eh* is stabilized at a low level and the soil pH is also stabilized near neutrality.

This chapter deals with the submerged soil system that has attained the latter stage of a steady state, and attempts to clarify the solubility and redox equilibria of the iron system in submerged soils in view of its importance in determining the soil *Eh* and pH. It is pertinent here to note the following:

- The soil *Eh*, even in a steady state, has little thermodynamic significance because of the heterogeneity of the system and difficulties inherent in measuring *Eh* such as slow electrode reaction and mixed potentials. The soil solution *Eh* and pH better represent the thermodynamic equilibrium state. Therefore, the theoretical considerations developed in this chapter are relevant to the soil solution phase.

- Since thermodynamic constants for the compounds occurring ephemerally in an amorphous state are rarely available, those available for better-defined compounds are used for the theoretical treatments. Therefore, the results of computations are inevitably approximations to the real condition. Consequently, the effect of ionic strength is not considered and the activity coefficient of relevant chemical species is assumed to be at unity.

6.2 pH shift in submerged soils

The rise of pH in the soil and the soil solution of originally acidic soils under submerged conditions has been known since the early days of paddy soil research. The most qualitative way to explain the rise in pH is to compare the substances in the oxidized and reduced conditions. Mn^{IV}, Fe^{III}, NO_3^-, SO_4^{2-} are stable in the oxidized condition and Mn^{II}, Fe^{II}, NH_4^+, and S^{2-} are stable in the reduced condition, the latter set of substances being more basic than the former.

More quantitative studies, paying attention to the apparent correlation between the Eh and pH, were conducted by, among others, Jeffery (1960) and Asami (1970). They found that when a soil is rich in reducible iron oxides, there is a highly significant correlation between the Eh and pH and that the dEh/dpH of the regression line seems to conform to that for the $Fe(OH)_3$-Fe^{2+} system, that is:

$$Eh = 1.06 - 0.059 \log (Fe^{2+}) - 0.177 \text{ pH}$$

They considered that this strongly suggests that the iron system is the potential determining system of reduced soils. However, there are many other similar studies whose results do not necessarily coincide with the above. For example, Kobo and Konno (1970) empirically found a value of -0.110 for dEh/dpH for the steady state soil Eh and pH after several weeks of submergence. Yamane and Sato (1970) found that the dEh/dpH value for field soils changes at different stages of soil reduction, indicating that redox systems other than the iron system would be operating.

Ponnamperuma (1972) explained the pH rise in terms of the iron system. As already seen, all the important reduction reactions occurring in nature involve the consumption of H^+ ions. This means a decrease in acidity or an increase in net OH^- ion concentration. The degree of acidity decrease or pH increase is determined by the H^+ ion consumption per mole of electron transfer, and it is largest for the reduction of $Fe(OH)_3$. Furthermore, as iron oxides are the most abundant oxidant in the soil, the rise of pH in acid soils is largely determined by the reduction of iron.

However, Ponnamperuma (1972), distinguished two stages at which different iron systems operate. Before the peak of water soluble iron formation, the $Fe(OH)_3$-Fe^{2+} system appears to explain the Eh-pH relationship. However, after the peak of iron dissolution, the following equations fit with the experimental Eh–pH relationship:

$$(1/2) \, Fe_3(OH)_8 + 4H^+ + e = (3/2) \, Fe^{2+} + 4H_2O$$
$$3Fe(OH)_3 + H^+ + e = Fe_3(OH)_8 + H_2O$$

Ponnamperuma (1972) related the rise in the pH of submerged soils primarily to the reduction of ferric iron, but he explained the fairly stable pH attained after a few weeks of submergence by the carbonate system. He found the following empirical relationship for reduced ferruginous soils:

$$pH = 6.1 - 0.58 \log P_{CO2}$$

This relationship is almost identical to that for an aqueous suspension of $Fe_3O_4 \cdot nH_2O$ equilibrated with CO_2 (Ponnamperuma et al., 1969).

Ponnamperuma (1972) further showed that calcareous and sodic soils with alkaline pH values before submergence lower their pH after reduction, and that their pH values are related to the $CaCO_3$-H_2O-CO_2 and Na_2CO_3-H_2O-CO_2 equilibrium, respectively.

The high sensitivity of the pH values (and therefore the Eh) of submerged soils, whether acid or alkaline, to carbonate concentration was shown by Ponnamperuma (1972), as in Figure 6.1. Regardless of the original soil pH, bubbling the solution with N_2 gas raised the soil solution pH of a submerged soil via the loss of CO_2.

Thus, it is necessary to discuss the dissolved carbon dioxide system and its effect on soil pH in the submerged condition.

6.3 Dissolved carbon dioxide in the soil solution

Soil organic matter is utilized by microorganisms as an energy source in aerobic respiration, anaerobic respiration and in fermentation, thereby releasing carbon dioxide as the final product of organic matter decomposition. As seen in Figure 5.3, the rapid emission of CO_2 occurs as the soil is reduced. Many studies conducted on the dissolved carbon dioxide in paddy soils show a range of C_T (total dissolved carbonate) between $10^{-1.5} \, M$ (or 717 ml CO_2 litre^{-1}) and $10^{-2} \, M$ (or 224 ml CO_2 litre^{-1}), or a partial pressure of CO_2 between 0.2 and 0.8 atm one to three weeks after submergence and between 0.05 and 0.2 atm during the later period. Of course, the kinetics of CO_2 evolution is dependent upon soil conditions, especially on the content of organic matter and iron oxide. However, CO_2 is usually formed very vigorously during the early period of submergence when aerobes and facultative anaerobes are active, while its evolution declines later as fermentation, especially CH_4 formation, begins.

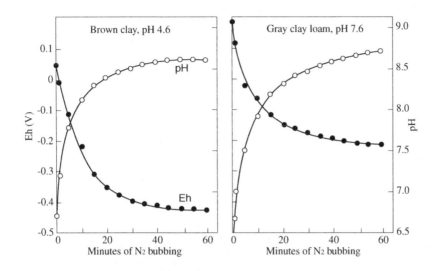

Figure 6.1
Influence of loss of CO_2 (caused by bubbling N_2) on pH and Eh of the solutions
of two soils (ten weeks after submergence) (Source: Ponnamperuma, 1972)

In this section, the carbonate equilibrium in the soil solution is considered after Stumm and Morgan (1970) in their treatise *Aquatic Chemistry*. In simple carbonate solutions, there are four species: $CO_2(aq)$, H_2CO_3, HCO_3^-, and CO_3^{2-}, of which the first two are conventionally expressed as:

$$[H_2CO_3^*] = [CO_2(aq)] + [H_2CO_3] \tag{1}$$

where [ ] denotes concentration. As it is estimated that $[CO_2(aq)]/[H_2CO_3]$ ≒ 650 at 25°C, the concentration of $CO_2(aq)$ is nearly identical to the analytical concentration of $H_2CO_3^*$.

From an analytical viewpoint, C_T is the easiest to determine, where

$$C_T = [H_2CO_3^*] + [HCO_3^-] + [CO_3^{2-}] \tag{2}$$

If a carbonate solution is acidified, HCO_3^- and CO_3^{2-} ions are transformed to H_2CO_3 and $CO_2(aq)$ and the solution becomes supersaturated with respect to $CO_2(g)$. Under reduced pressure, $CO_2(g)$ is collected quantitatively and measured for C_T.

C_T is determined by the solubility equilibrium:

$$CO_2(g) \rightleftharpoons CO_2(aq)$$

If CO_2 gas is assumed to behave ideally and $[CO_2(aq)] = [H_2CO_3^*]$, the distribution constant K_D is written as:

$$K_D = [CO_2(aq)] / [CO_2(g)]$$

Furthermore, $CO_2(g)$ is expressed using Dalton's law of partial pressure as:

$$[CO_2(g)] = P_{CO2}/RT$$

Combining the above two

$$[CO_2(aq)] = (K_D/RT) P_{CO2}$$
$$= K_H P_{CO2} \tag{3}$$

where $K_H = K_D/RT$ (mole litre^{-1} atm^{-1}), which is the Henry's law constant and is equal to $10^{-1.5}$ at 25°C.

When the medium pH is high, dissolved $CO_2(aq)$ or $H_2CO_3^*$ further dissociates to produce HCO_3^- and CO_3^{2-}, which also contribute to C_T. Thus:

$$K_1 = [H^+] [HCO_3^-] / [H_2CO_3^*] = 10^{-6.4} \tag{4}$$
$$K_2 = [H^+] [CO_3^{2-}] / [HCO_3^-] = 10^{-10.3}. \tag{5}$$

The ionization fractions for each carbonate species, $\alpha_0 \sim \alpha_2$, are defined as follows:

$$[H_2CO_3^*] = C_T\alpha_0$$

$$[HCO_3^-] = C_T\alpha_1$$

$$[CO_3^{2-}] = C_T\alpha_2$$

wherein $\alpha_0 + \alpha_1 + \alpha_2 = 1$.
From equations (4) and (5):

$$[HCO_3^-] = K_1[H_2CO_3^*]/[H^+]$$
$$[CO_3^{2-}] = K_1K_2[H_2CO_3^*]/[H^+]^2$$

Hence, from (2):

$$C_T = [H_2CO_3^*](1 + K_1/[H^+] + K_1K_2/[H^+]^2) \tag{6}$$

and so:

$$\alpha_0 = (1 + K_1/[H^+] + K_1K_2/[H^+]^2)^{-1} \tag{7}$$

Likewise:

$$\alpha_1 = ([H^+]/K_1 + 1 + K_2/[H^+])^{-1} \tag{8}$$
$$\alpha_2 = ([H^+]^2/K_1K_2 + [H^+]/K_2 + 1)^{-1} \tag{9}$$

Accordingly:

$$[H_2CO_3^*] = C_T[H^+]^2/([H^+]^2 + K_1[H^+] + K_1K_2) \tag{10}$$
$$[HCO_3^-] = C_TK_1[H^+]/([H^+]^2 + K_1[H^+] + K_1K_2) \tag{11}$$
$$[CO_3^{2-}] = C_TK_1K_2/([H^+]^2 + K_1[H^+] + K_1K_2). \tag{12}$$

When $(H^+) > K_1 > K_2$ or $pH < pK_1 < pK_2$, where $pK_1 = -\log K_1$ and $pK_2 = -\log K_2$:

$$\log [H_2CO_3^*] = \log C_T$$
$$\log [HCO_3^-] = \log C_T + \log K_1 + pH$$
$$\log [CO_3^{2-}] = \log C_T + \log K_1 + \log K_2 + 2pH$$

When $pK_1 < pH < pK_2$:

$$\log [H_2CO_3^*] = \log C_T - \log K_1 - pH$$
$$\log [HCO_3^-] = \log C_T$$
$$\log [CO_3^{2-}] = \log C_T + \log K_2 + pH$$

When $pK_1 < pK_2 < pH$:

$$\log [H_2CO_3^*] = \log C_T - \log K_1 - \log K_2 - 2pH$$
$$\log [HCO_3^-] = \log C_T - \log K_2 - pH$$
$$\log [CO_3^{2-}] = \log C_T$$

Figure 6.2 illustrates the aforementioned relationships.

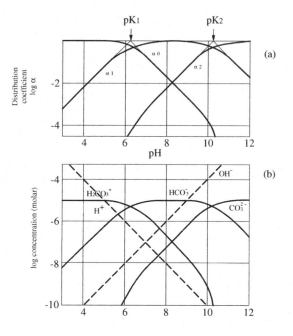

Figure 6.2
*Distribution of solute species in aqueous carbonate systems: (a) ionization
fractions as a function of pH, (b) logarithmic equilibrium diagram for fresh water
under atmospheric CO_2 (Source: Stumm and Morgan, 1970, partly modified)*

As $[CO_2(aq)] = K_H \cdot P_{CO2}$, and at the same time $[CO_2(aq)] \approx [H_2CO_3^*]$, the
relationship between C_T and P_{CO2} is written, according to the equation (6),
as :

$$C_T = K_H \cdot P_{CO2} (1 + K_1/[H^+] + K_1 K_2/[H^+]^2) \qquad (13)$$

or

$$P_{CO2} = C_T / [K_H(1 + K_1/[H^+] + K_1 K_2/[H^+]^2)] \qquad (14)$$

Thus, for the same P_{CO2} value, C_T varies with the pH of the solution.
Figure 6.3(a) shows the aqueous carbonate equilibrium at a constant P_{CO2}
of $10^{-3.5}$ atm, namely, a solution in equilibrium with the atmospheric CO_2.
A strong acid or a strong base is used to alter the pH. As seen in this figure,
C_T is constant in the pH range where $H_2CO_3^*$ is predominant; that is, the pH
is below pK_1. It increases with a slope of +1 in the pH range between pK_1

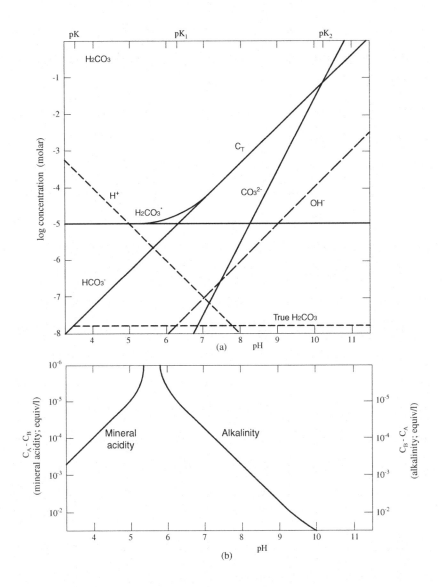

Figure 6.3
Aqueous carbonate equilibrium: (a) water is equilibrated with the atomosphere and the pH is adjusted with strong base or strong acid; (b) alkalinity or mineral acidity at pH values different from that of a pure CO_2 solution.
(Source: Stumm and Morgan, 1970)

and pK_2, and with a slope of $+2$ in the pH range higher than pK_2. In Figure 6.3(a), equilibrium concentrations under the atmospheric partial pressure of CO_2 read as:

$$- \log [H^+] = - \log [HCO_3^-] = 5.65^{(1)}$$
$$- \log [CO_2(aq)] = - \log [H_2CO_3^*] = 5.0$$
$$- \log [CO_3^{2-}] = 10.4 \text{ (at pH} = 5.65)$$

A pure CO_2 solution has the following charge balance:

$$[H^+] = [HCO_3^-] + 2[CO_3^{2-}] + [OH^-] \tag{15}$$

At pH values different from that of a pure CO_2 solution, the solution contains either alkalinity or mineral acidity, depending on the concentration of the strong acid C_A or strong base C_B that was to be added to achieve these pH values. Considering that C_B is equivalent to the concentration of a monovalent cation (other than H^+, such as Na^+ from NaOH) and that C_A is equivalent to the concentration of a monovalent anion (other than OH^-, such as Cl^- from HCl), the charge balance is expressed as:

$$C_B + [H^+] = [HCO_3^-] + 2[CO_3^{2-}] + [OH^-] + C_A$$
$$C_B - C_A = C_T(\alpha_1 + 2\alpha_2) + [OH^-] - [H^+]$$
$$= K_H \cdot P_{CO2}(\alpha_1 + 2\alpha_2)/\alpha_0 + [OH^-] - [H^+]$$
$$= K_H \cdot P_{CO2}(K_1[H^+] + 2K_1K_2)/[H^+]^2 + [OH^-] - [H^+] \tag{16}$$
$$= [\text{Alkalinity}] \text{ (equivalent or mol}_c \text{ litre}^{-1})$$

$(C_B - C_A)$ is the acid neutralizing capacity of the solution with respect to the pure solution of CO_2 and thus generates the alkalinity, [Alk]. If $C_A > C_B$, then $(C_A - C_B)$ represents the base neutralizing capacity or the mineral acidity. Figure 6.3(b) shows the relationship between the alkalinity and pH or mineral acidity and pH when the carbonate solution is in equilibrium with the atmosphere.

6.4 The solubility equilibrium of ferrous iron in the carbonate system

The more realistic conditions for actual paddy soils need to be considered. Under the constant P_{CO2} of $10^{-1.5}$ atm, the equilibrium pH is about 4.7 and $[H_2CO_3^*] = 10^{-3} M$. If a soil solution in equilibrium with $10^{-1.5}$ atm P_{CO2} has a pH of 7, the alkalinity should be $10^{-2.4}$ equivalent litre^{-1}, which is required to shift the equilibrium pH from 4.7 to pH 7, according to equation (16).

What then makes up this alkalinity? Ferrous iron that is produced upon reduction should be the most important cation that sustains the alkalinity. If a soil does not contain a sufficient amount of reducible ferric iron, even when the reduction proceeds and a high amount of carbonate is produced, the soil pH does not increase to 6.5–7. This fact suggests that the pH increase in normal (or ferruginous) submerged soils is due to an accumulation of Fe^{2+} that sustains a high concentration of bicarbonate in soil solution as alkalinity. Of course, all other mono- and divalent cations that are dissolved in a soil solution in the process of reduction or through cation exchange for Fe^{2+} should also sustain the alkalinity. According to previous studies, Ca^{2+} and Mg^{2+} are the most important cations, besides Fe^{2+}, present in the soil solution of submerged paddy soils. Therefore, the sum of cations that sustains the alkalinity of the soil solution is expressed as M^{2+}, which consists mainly of $Fe^{2+} + Ca^{2+} + Mg^{2+}$. The charge balance equation may be written as follows:

$$2[M^{2+}] + [H^+] = [HCO_3^-] + 2[CO_3^{2-}] + [OH^-] \qquad (17)$$
$$[M^{2+}] = K_H \cdot P_{CO2} (K_1[H^+] + 2K_1K_2)/(2[H^+]^2) + ([OH^-] - [H^+])/2 \quad (18)$$

When pH $< pK_2$ and near neutrality:

$$\log [M^{2+}] = -8.2 + \log P_{CO2} + pH \qquad (19)$$

The amount of Fe^{2+} that can be present in the soil solution needs to be considered in more detail, because Fe^{2+} solubility is presumably governed by either of the following solubility products (under normal conditions sulfide is not considered):

$Fe(OH)_2$	$K_{spH} = 10^{-14.7}$
$FeCO_3$	$K_{spC} = 10^{-10.7}$
$Fe_3(OH)_8$	$K_{spF} = 10^{-17.6}$

For the coexistence of the Fe^{2+}-$Fe(OH)_2$ system and the Fe^{2+}-$FeCO_3$ system, the following must be held simultaneously:

	$\log K$
$Fe^{2+} + 2OH = Fe(OH)_2(s)$	14.7
$FeCO_3(s) = Fe^{2+} + CO_3^{2-}$	−10.7
$FeCO_3(s) + 2OH^- = Fe(OH)_2(s) + CO_3^{2-}$	4.0

Thus:

$$\log [CO_3^{2-}] - 2\log [OH^-] = 4.0 \tag{20}$$

From the second dissociation constant of carbonic acid:

$$[CO_3^{2-}] = K_1 K_2 K_H P_{CO2}/[H^+]^2$$

When the numerical values are put into K_1, K_2, and K_H in the above equation:

$$\log [CO_3^{2-}] = -18.2 + \log P_{CO2} + 2pH \tag{21}$$

Substituting this for $\log [CO_3^{2-}]$ in the equation (20), then $\log P_{CO2} = -5.8$. This means that at $P_{CO2} = 10^{-5.8}$ atm, $Fe(OH)_2$ and $FeCO_3$ can coexist as the stable solid phase, but at a higher P_{CO2}, $FeCO_3$ becomes the sole stable solid phase. In the natural soil system, the equilibrium P_{CO2} cannot be lower than that of the atmosphere, so $FeCO_3$ is the stable solid phase that determines Fe^{2+} solubility. It is considered virtually impossible for $Fe(OH)_2(s)$ to form in soils (Lindsay, 1979).

Similarly, the condition for the coexistence of the Fe^{2+}-$FeCO_3$ system and the Fe^{2+}-$Fe_3(OH)_8$ system should be sought:

$$
\begin{array}{lr}
 & \log K \\
2Fe(OH)_3(s) + Fe^{2+} + 2OH^- = Fe_3(OH)_8(s) & 17.6 \\
FeCO_3(s) = Fe^{2+} + CO_3^{2-} & -10.7 \\
\hline
2Fe(OH)_3(s) + FeCO_3(s) + 2OH^- = Fe_3(OH)_8(s) + CO_3^{2-} & 6.9
\end{array}
$$

$$\log [CO_3^{2-}] - 2\log [OH^-] = 6.9 \tag{22}$$

Substituting (21) for $\log [CO_3^{2-}]$ in (22), then $\log P_{CO2} = -2.9$. Therefore, when $P_{CO2} > 10^{-2.9}$ atm, $FeCO_3$ is more stable than $Fe_3(OH)_8$, whereas at lower P_{CO2} (between 1.15×10^{-3} and 3×10^{-4} atm) the latter is a dominant solid species in equilibrium with Fe^{2+}.

In submerged soils, P_{CO2} is reported to be 10 to 100 times that of the atmosphere, or from $10^{-2.5}$ to $10^{-1.5}$ atm. Therefore, it is reasonable to assume the presence of $FeCO_3$ as the solid phase that is in equilibrium with Fe^{2+} in the soil solution. Ponnamperuma (1972) considers that $Fe_3(OH)_8$ is the stable solid in the redox equilibrium of iron oxide hydrates in ordinary submerged soils, as ferrous carbonate takes weeks to precipitate from supersaturated solutions. This latter reasoning, does not exclude the possibility of $FeCO_3$ being the stable solid phase in the steady state of reduction that is attained after a few to several weeks of submergence. Moreover, Lindsay (1979) considers that as P_{CO2} rises above 0.03 atm, hematite and goethite can transform directly to siderite without first forming ferrosic oxide. It is also an established fact that $FeCO_3$ or siderite is present as white concretions or mottles in strongly reduced paddy soils (Yamazaki and Yoshizawa, 1961).

Assuming $FeCO_3(s)$ to be the stable solid phase, the solubility of Fe^{2+} is expressed, using equation (21), as:

$$\log [Fe^{2+}] = -10.7 - \log [CO_3^{2-}]$$
$$= -10.7 - (-18.2 + \log P_{CO2} + 2pH)$$
$$= 7.5 - \log P_{CO2} - 2pH \qquad (23)$$

In the $FeCO_3$-H_2O-CO_2 system, the charge balance equation (19) and the solubility relationship (23) must be held simultaneously:

$$\log [M^{2+}] = -8.2 + \log P_{CO2} + pH$$
$$\log [Fe^{2+}] = 7.5 - \log P_{CO2} - 2pH$$

Here, a parameter, must be assumed that gives the fraction of $[Fe^{2+}]$ in $[M^{2+}]$, so the charge balance equation can be expressed in terms of $[Fe^{2+}]$. It is necessary to define $[M^{2+}]/[Fe^{2+}] = \beta$, which is a parameter related to the CEC and the base status of a soil. The charge balance equation can be rewritten as

$$\log [Fe^{2+}] = -8.2 - \log\beta + \log P_{CO2} + pH \qquad (24)$$

When this is equated with the solubility equation:

$$pH = 5.23 + (1/3)\log\beta - (2/3) \log P_{CO2} \qquad (25)$$
$$\log P_{CO2} = 7.85 + (1/2)\log\beta - (3/2) pH \qquad (26)$$

Substituting equation (25) for the pH in equation (24), then:

$$\log [Fe^{2+}] = -2.97 - (2/3)\log\beta + (1/3)\log P_{CO2} \qquad (27)$$

Thus, if both the partial pressure of carbon dioxide gas in equilibrium with the soil solution and the parameter β are given, the pH of the medium and the solubility of Fe^{2+} are determined.

An extreme case of a high base status or a high value of β is a calcareous soil in which free $CaCO_3$ is present. When such a soil is submerged and reduced, $CaCO_3$ and $FeCO_3$ should exist as the solid phase, and this determines the $[Ca^{2+}]/[Fe^{2+}]$ ratio in the soil solution according to the respective solubility products, as follows:

$$[Ca^{2+}]/[Fe^{2+}] = [Ca^{2+}][CO_3^{2-}]/[Fe^{2+}][CO_3^{2-}]$$
$$= 10^{-8.35}/10^{-10.7} = 10^{2.35}$$

Thus:

$$\log [Ca^{2+}] = 2.35 + \log [Fe^{2+}] \tag{28}$$

As $[Ca^{2+}]$ predominates over other cations in such a soil, the charge balance equation (19) may be written as:

$$\log [Ca^{2+}] = -8.2 + \log P_{CO2} + pH \tag{29}$$

The solubility of Ca^{2+} as a function of P_{CO2} and pH can also be derived from the solubility product equation and equation (21):

$$\log [Ca^{2+}] = -8.35 - \log [CO_3^{2-}]$$
$$= 9.85 - \log P_{CO2} - 2pH \tag{30}$$

Equating (29) with (30), then:

$$pH = 6.02 - (2/3) \log P_{CO2} \tag{31}$$

The coexisting Fe^{2+} concentration is given as a function of P_{CO2} as follows:

$$\log [Fe^{2+}] = -4.53 + (1/3) \log P_{CO2} \tag{32}$$

This has been derived from (28), (29) and (31).

Since $[Fe^{2+}]$ is only about 1/220 of $[Ca^{2+}]$, it can barely affect the charge balance or the equilibrium pH of the system. In other words, the redox equilibrium of the iron system has little to do with the pH of the calcareous soil. When P_{CO2} ranges from $10^{-1.5}$ to $10^{-2.5}$ atm, the expected pH of a submerged soil ranges between 7 and 7.7, whereas an upland calcareous soil in equilibrium with atmospheric CO_2 ($P_{CO2}=10^{-3.5}$atm) has a pH of about 8.4[2].

6.5 The redox equilibrium of the $Fe(OH)_3$-Fe^{2+}-$FeCO_3$ system

Next, the redox equilibrium of submerged soils is considered, in relation to the solubility equilibrium of ferrous iron. If the $Fe(OH)_3$-Fe^{2+} system is considered to be the potential determining system, the redox reaction should follow the equation:

$$Fe(OH)_3 + 3H^+ + e = Fe^{2+} + 3H_2O \qquad \begin{array}{c} \log K \\ 17.9 \end{array} \qquad (33)$$

Combining this with the solubility product equation for $FeCO_3$, the following redox half reaction is obtained:

$$Fe(OH)_3 + 3H^+ + CO_3^{2-} + e = FeCO_3 + 3H_2O \qquad 28.6 \qquad (34)$$

Knowing pE° to be 28.6, the redox potential for the above reaction can be written, substituting (21) for $\log [CO_3^{2-}]$, as:

$$\begin{aligned} Eh &= 1.69 + 0.059\log [CO_3^{2-}] - 0.177 \text{ pH} \\ &= 0.61 + 0.059 \log P_{CO2} - 0.059 \text{ pH} \end{aligned} \qquad (35)$$

Furthermore, the pH in the equation may be substituted by the previously obtained equation (25), that is:

$$\text{pH} = 5.23 + (1/3) \log\beta - (2/3)\log P_{CO2}$$

to produce an Eh equation as a function of the base status parameter and P_{CO2}:

$$Eh = 0.30 - 0.02 \log\beta + 0.098\log P_{CO2} \qquad (36)$$

Or, if P_{CO2} is expressed as a function of pH, as in equation (26):

$$\log P_{CO2} = 7.85 + (1/2)\log\beta - (3/2)\text{pH}$$

then:

$$Eh = 1.07 + 0.03\log\beta - 0.148 \text{ pH} \qquad (37)$$

This equation illustrates that the $dEh/d\text{pH} = -0.148$ volt for the assumed iron system. There have been many arguments on the $dEh/d\text{pH}$

value both from empirical and theoretical viewpoints. The most widely accepted value is −59 mV, but some authors adopt −177 mV. The value obtained here, −148 mV, is equal to −2.5 × 2.3 RT/F and is close to some experimental values. Further experimental verifications is necessary.

6.6 Ranges of log [Fe²⁺], pH, and *Eh* as a function of P_{CO2} as the master variable

As demonstrated above, log [Fe²⁺], pH, and *Eh* can be calculated as a function of P_{CO2} according to the following equations:

$$\log [Fe^{2+}] = -2.97 - (2/3)\log\beta + (1/3)\log P_{CO2} \tag{27}$$
$$pH = 5.23 + (1/3)\log\beta - (2/3)\log P_{CO2} \tag{25}$$
$$Eh = 0.30 - 0.02\log\beta + 0.098\log P_{CO2} \tag{36}$$

whereby it is assumed that:
- The solubility of Fe²⁺ is determined by the solubility product of FeCO₃;
- The redox potential is determined by the Fe(OH)₃-Fe²⁺-FeCO₃ system.

The range of the master variable, P_{CO2}, should be set at 10 times to 1000 times that of the atmosphere, that is 0.003 or $10^{-2.5}$ atm to 0.3 or $10^{-0.5}$ atm, considering the experimental findings. As previously discussed, in this range of P_{CO2}, FeCO₃ is the sole thermodynamically stable solid phase that determines the solubility of ferrous iron.

Another parameter ought to be assumed, that is [M²⁺]/[Fe²⁺] = β. This parameter is related to the base status of the soil (CEC, percentage base saturation, and exchangeable cation composition). The value of β would normally vary from 1 for completely unsaturated acid soils to $10^{2.35}$ for calcareous soils; these two extremes respectively set the lower and upper limits of pH to be attained by reduction. The following calculation assumes the values of β to be 1, 2, 5, and 10 for non-calcareous soils and $10^{2.35}$ for calcareous soils. The result of the computation is given in Table 6.1.

The computed [Fe²⁺] ranges from $10^{-3.11}$ to $10^{-5.4}$ *M*, depending on P_{CO2} and β, but in the normal range of these two variables for acid soils, it seems to vary from $10^{-3.5}$ to $10^{-4.0}$ *M*, or ca. 5 to 15 mg Fe²⁺ litre⁻¹. The estimated values seem to be too low when compared with the experimentally obtained Fe²⁺ concentration, a few hundred to several hundred mg litre⁻¹. Two possible explanations are:
- A higher solubility of ephemerally precipitated amorphous FeCO₃ than crystalline siderite for which the solubility product is available;

- The high experimental Fe^{2+} concentration may be partly due to undissociated species, such as the one chelated by organic ligands and ion pairs (such as $FeCO_3^0$, $FeHCO_3^+$).

The computed pH varies from 5.5 to 7.7, but in the normal P_{CO2} and β ranges for acid soils it varies from 5.7 to 6.5. The experimentally measured pH range for submerged acid soils is pH 6–7. The computed pH seems to be a little lower and a possible explanation for the discrepancy would be:

- The normally reported pH is the soil pH, but not the solution pH. It is known that the solution pH at the steady state is lower than the soil pH by 0.3–0.5 pH units (Sato and Yamane, 1973);
- The experimental soil pH may be higher than the true soil pH, because the experimental procedure may cause a loss of dissolved CO_2;
- If Fe^{2+} solubility is 10 times higher than assumed, because of a higher solubility product of amorphous $FeCO_3$, the pH should be 1/3 unit higher[3] than that computed.

Table 6.1 Ranges of $log[Fe^{2+}]$, pH and Eh for steady state soil solution of submerged soils, computed as functions of master variable P_{CO2} and base status parameter, β

Base status parameter, β	log P_{CO2}, atm				
	−2.5	−2.0	−1.5	−1.0	−0.5
	log [Fe²⁺] (Fe²⁺ in *M*)				
1	−3.83	−3.67	−3.50	−3.33	−3.11
2	−3.99	−3.83	−3.66	−3.49	−3.33
5	−4.26	−4.10	−3.93	−3.76	−3.60
10	−4.47	−4.31	−4.14	−3.97	−3.79
10E2.35*	−5.36	−5.20	−5.03	−4.86	−4.70
	pH				
1	6.90	6.56	6.23	5.90	5.53
2	7.00	6.66	6.33	6.00	5.66
5	7.14	6.80	6.47	6.14	5.80
10	7.24	6.90	6.57	6.24	5.90
10E2.35*	7.69	7.35	7.09	6.69	6.35
	Eh (in V)				
1	0.058	0.107	0.156	0.205	0.254
2	0.053	0.102	0.151	0.200	0.249
5	0.044	0.093	0.142	0.191	0.240
10	0.038	0.087	0.136	0.185	0.234
10E2.35*	0.012	0.061	0.110	0.159	0.208

* 10 to the 2.35 power, assuming existence of free calcium carbonate

The computed Eh value varies from 100 to 250 mV in the normal range of P_{CO2} and β for acid soils. These values are much higher than the reported value range of 0 to -200 mV. Plausible explanations for the discrepancy would be:

- The reported steady state Eh value is again the soil Eh, but not the solution Eh. Sato and Yamane (1973) found that the solution Eh is about 300 mV higher than the corresponding soil Eh;
- Correction for a higher Fe^{2+} solubility lowers the Eh. A ten times increase in Fe^{2+} solubility lowers the Eh by about 80 mV[4].

The computed values of log $[Fe^{2+}]$, pH, and the Eh for calcareous soils or the β value of $10^{2.35}$ indicate that in the presence of free $CaCO_3$, the Fe^{2+} concentration is lowered, the pH is raised, and the Eh is lowered compared with non-calcareous acid soils. This is of interest from the practical point of view in relation to liming of paddy soils to counter '*bronzing*' or iron toxicity of rice. Although the cause of this physiological disorder seems to be complex, liming at least would lower the Fe^{2+} concentration very significantly.

Notes

(1) The same result was already obtained in Note (3) of Chapter 4.

(2) The pH of a soil containing free $CaCO_3$ may also be calculated as follows:

Calcium carbonate in the soil is dissolved by carbonic acid

$$CaCO_3 + H_2CO_3 = Ca^{2+} + 2HCO_3^- \qquad \text{(i)}$$

The equilibrium constant of this reaction, K, is written as follows:

$$
\begin{aligned}
K &= [Ca^{2+}][HCO_3^-]^2 / [H_2CO_3] \qquad \text{(ii)}\\
&= ([Ca^{2+}][CO_3^{2-}])\cdot([HCO_3^-][H^+]/[H_2CO_3])\cdot([HCO_3^-]/[CO_3^{2-}][H^+])\\
&= K_{sp}\cdot K_1 \cdot K_2^{-1} = 10^{-4.4}
\end{aligned}
$$

where $K_{sp} = 10^{-8.3}$ is the solubility product of $CaCO_3$, $K_1 = 10^{-6.4}$ and $K_2 = 10^{-10.3}$ are the first and second dissociation constants of carbonic acid, respectively.

Now, postulating that the dissolved carbonate is in equilibrium with atmospheric CO_2 (0.03%), the carbonate concentration is:

$$[H_2CO_3^*] = K_H P_{CO2} = 10^{-5} M.$$

As it is known from equation (i) that $[Ca^{2+}] = 1/2[HCO_3^-]$, equation (ii) may be rewritten as:

$$K = [HCO_3^-]^3 / (2 \times 10^{-5}) = 10^{-4.4}$$

so: $[HCO_3^-] = 10^{-3.03} M$
and: $[Ca^{2+}] = 10^{-3.33} = 4.7 \times 10^{-4} M$
Therefore, the pH of the soil solution can be calculated from the equation for K_1:

$$[H^+] = 10^{-8.4} \text{ or pH} = 8.4$$

This is the reason why saline soils are considered to have a pH < 8.5, as they almost always contain free calcium carbonate.

(3) If $\log K_{spC} = -9.7$ instead of -10.7, then:

$$\log [Fe^{2+}] = 8.5 - \log P_{CO2} - 2pH$$

and the equation for the pH is:

$$pH = 5.56 + (1/3) \log\beta - (2/3) \log P_{CO2}$$

(4) If $\log K_{spC} = -9.7$, the log K value for the reaction:

$$Fe(OH)_3 + 3H^+ + CO_3^{2-} + e = FeCO_3 + 3H_2O$$

would be 27.6 instead of 28.6. Then the pE° should be lower by 1 or E° lower by 59 mV. In addition, the higher pH by 1/3 unit should lower another 20 mV according to the equation (35).

References

Asami, T. 1970. Behavior of free iron oxide in paddy soils (Part 3). On the relationship between ferrous iron, *Eh* and pH in paddy soils. *J. Sci. Soil & Manure, Japan*, 41: 45–47. (In Japanese).

Jeffery, J.W.O. 1960. Iron and the Eh of waterlogged soils with particular reference to paddy. *J. Soil Sci.*, 11: 140–148.

Kobo, K. and Konno, T. 1970. Studies on the transformation of the materials in paddy soils under leaching condition (Part 1). Changes of pH and

Eh, occurrence of CO_2 and Fe(II). *J. Sci. Soil & Manure, Japan*, 41: 178–187. (In Japanese).

Lindsay, W.L. 1979. *Chemical Equilibria in Soils*. John Wiley & Sons, N.Y.

Ponnamperuma, F.N. 1972. The chemistry of submerged soils. *Adv. in Agron.*, 24: 29–96.

Ponnamperuma, F.N. Castro, R.U. and Valencia, C.M. 1969. Experimental study of the influence of the partial pressure of carbon dioxide on the pH values of aqueous carbonate systems. *Soil Sci. Soc. Amer. Proc.*, 33: 239–241.

Sato, K. and Yamane, I. 1973. Studies on soil solution of submerged soil (Part 2). Composition of soil solution. *J. Sci. Soil & Manure, Japan*, 44: 246–250. (In Japanese).

Stumm, W. and Morgan, J.J. 1970. *Aquatic Chemistry: An Introduction Emphasizing Chemical Equilibria in Natural Waters*. John Wiley & Sons, N.Y.

Yamane, I. and Sato, K., 1970. Plant and soil in a lowland rice field added with forage residues. *Rep. Inst. Agric. Res. Tohoku Univ.*, 21: 79–101.

Yamazaki, T. and Yoshizawa, T. 1961. Mottles and concretions of ferrous carbonate ($FeCO_3$) occurring in paddy soil profiles, Part 1. *Bull. Hokuriku Agric. Expt. Sta.*, 2: 1–16. (In Japanese).

Chapter 7: Long-Term Chemical and Morphological Changes Induced by Alternating Submergence and Drainage of Paddy Soils

Some chemical and morphological characteristics of paddy soils are the result of processes induced by alternating submergence and drainage. Some of the changes are transient, while others leave permanent marks on the soil. This chapter reviews the processes of changes induced by management practices that are specific to paddy soils.

7.1 Accumulation of soil organic matter

Paddy soils are submerged, either artificially or naturally, for at least three to four months a year. During most of this period, the soil is kept under a reduced condition with a greatly restricted oxygen supply. Therefore, decomposition of organic matter is severely inhibited, and fermentation becomes the main form of its transformation. As a result, the amount of organic matter increases in paddy soils, compared with upland soils. Figure 7.1 clearly shows the long-term effect of submergence on the accumulation of soil organic matter (Mitsuchi, 1974 c). Mitsuchi compared several sets of adjacent soils that differed only in their land use; one being under paddy rice cultivation and the other being kept for upland cultivation for many years.

Except for one case of ando soil (Udands), the soils used for paddy rice cultivation clearly show a higher organic matter content than upland soils— sometimes more than twice as high. One of the reasons for this may be a higher supply of organic matter in paddy soils. The total production of above-ground biomass in paddy soils in Japan is much higher (more than 10 t ha^{-1}) than upland soils growing either cereals or vegetables. Roots and stubble remaining in the soil after the harvesting of paddy fields would add another 2–3 t ha^{-1}. In addition, weeds growing during the drained period make a significant contribution to soil organic matter in Japan (Yamazaki and Saeki,

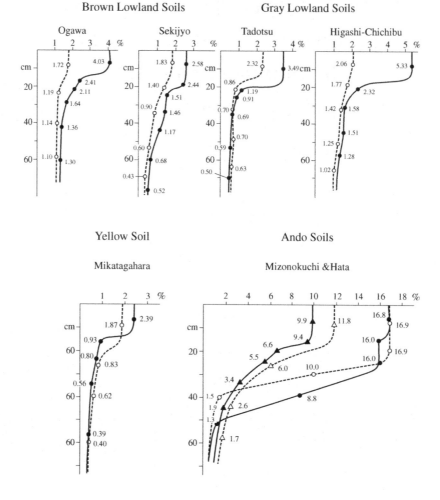

Figure.7.1

Comparison of organic matter content between paddy soils and adjacent upland soils (—— paddy soil: ---- upland soil) (Source: Mitsuchi, 1974c)

1980). All of these weeds are incorporated into the soil when preparing land for the subsequent crop, adding 2–3 t ha^{-1} of dry matter, equivalent to almost 12 t ha^{-1} (fresh weight) of manure or 23 t ha^{-1} (fresh weight) of green manure (Chinese milk vetch).

The inefficiency of decomposition under submergence also contributes to the accumulation of organic matter. Kanazawa and Yoneyama (1980) found that fungi are the major decomposers of coarse organic debris in upland soils. Fungi have a considerable ability to break down the skeletal components of plant debris, such as cellulose and lignin, which normally decompose with difficulty. Conversely, under the reduced conditions of paddy soils, aerobic microorganisms like fungi cannot grow well, and flourishing anaerobic bacteria generally cannot decompose these skeletal components.

Thus, organic matter in paddy soils is generally less well decomposed and less humified. Mitsuchi (1974c) illustrated this in his studies of the nature of soil humic acids for paddy and upland soils. Without exception, the two indices of humification (RF for the relative color depth and Δlog K for the slope of absorption spectrum of humic acid solution) for these soils, showed a shift towards a less humified nature of soil humus from upland to paddy soils.

Therefore, organic matter that accumulates in paddy soils is used as a source of nitrogen and other mineral nutrients. The mineralization of nitrogen from this source is considered important for rice nutrition, as is explained in Chapter 8.

7.2 Ferrolysis

The changes in the exchangeable cation composition of paddy soils, as demonstrated by the work of Kawaguchi and Kawachi (1969 a, b) in Chapter 3, should have a major implication for paddy soil characteristics, when the change has been accumulated by repetitive submergence and drainage.

Brinkman (1970) noted the importance of the same phenomenon of changes in exchangeable cations and cation exchange capacity induced by ferrous-ferric alternation, and called it 'ferrolysis' as a specific soil-forming process in seasonally submerged soils, of which paddy soil is a typical example. The process of ferrolysis leads to the formation of acid soils with pale surface horizons that are low in organic matter and clay. These soils are also low in bases and high in exchangeable aluminum, and have an anomalously low CEC due to the interlayering of 2:1 type clays with aluminum polymers and ferrous iron. These soils occupy large areas in Southeast Asia, mainly on the nearly level land of terrace landforms that are seasonally flooded by rainwater. They have a low natural fertility. Their physical properties are unfavorable for upland crops, but well suited for wetland rice. They contain more clayey subsoil material that creates an impervious layer for water percolation, so causing seasonal water stagnation and even submergence (Brinkman, 1978).

Brinkman's argument on the genesis of a soil that has undergone ferrolysis may be summarized as follows (Kyuma, 1978):

Redox condition: Reducing $\rightarrow$ Oxidizing $\rightarrow$ Reducing

$$\text{Clay transformation:} \quad Fe^{2+}-\text{clay} \rightarrow \begin{array}{c} H^+-\text{clay} \\ \downarrow \\ Al^{3+}-\text{clay} \end{array} \Bigg) \rightarrow \begin{array}{l} Fe^{2+}-\text{clay} \\ \text{with partial} \\ \text{interlayering} \end{array}$$

Soil reaction: Neutral $\rightarrow$ Acid $\rightarrow$ Neutral

Yoshida and Itoh (1974) conducted an experiment to confirm the process and came to the following conclusions:

- The degree of acidification is determined primarily by the degree of Fe^{2+} saturation in the exchange complex.
- Whether $Fe^{2+} \rightarrow H^+ \rightarrow Al^{3+}$ replacement occurs depends on the acid strength of the exchange sites, so Al^{3+} does not appear on allophane.

They also postulated the possibility of Al-interlayering of expanding 2:1 type clays, leading to a lowering of the CEC (see Table 3.3).

Mitsuchi (1974 b) compared paddy soils with adjacent upland soils, both of which had been derived from the same parent material containing vermiculite, and found a higher degree of Al-interlayering in the paddy soil.

Later, Brinkman (1977) presented more detailed evidence of Al-interlayering of expanding 2:1 type clays as a result of ferrolysis, whereby ferrous ions were trapped in the interlayer space.

The process called ferrolysis by Brinkman, which is the result of long-term cyclic alternation of submergence and drainage, appears to be important in paddy soils, inducing acidification and degradation on the one hand, and chloritization of expanding 2:1 type clay minerals on the other. In an advanced stage of ferrolysis on an old land surface, an albic horizon may form due to clay destruction, eventually accompanied by tonguing and interfingering of albic material into the underlying, more clayey horizon. A typical example is seen in the *Chhiata* soil on the Pleistocene terrace (Madhpur Tract) in Bangladesh (Brinkman, 1977). Soils at such an advanced stage of development may be classified at the great group level as, for example, an Albaquult. However, the formation of an albic horizon has rarely been reported in paddy soils, particularly in the recent sedimentary formations in which the majority of paddy soils are distributed. Moreover, ferrolysis can manifest itself clearly only in soil systems with 2:1 lattice clays. Thus, it is still premature to consider the classification of paddy soils that strongly exhibit ferrolysis.

7.3 Acidification and deacidification

Ferrolysis is just one mechanism for acidification of paddy soils. When soil contains large amounts of sulfidic material, acidification is obvious as, in the case of acid sulfate soils (see Chapter 11).

In the coastal plains, deltas, and polder lands, there are many paddy soils containing small amounts of sulfides or sulfates that originate from brackish sediments or are supplied from brackish ground-water. Under natural or artificial seasonal flooding and drying, as in the case of irrigated paddy fields, soils are known to acquire a pH of approximately 4.5 to 5 (Kubota, 1961; van Breemen, 1975; also see Chapter 2).

Van Breemen explained the process leading to the convergence of soil pH as follows:

When the soil pH is above 5, sulfate reducers are active and if the soil contains a small amount of SO_4^{2-}, the following reaction should occur during the submerged phase:

$$Fe\,(OH)_3 + SO_4^{2-} + (9/4)CH_2O = FeS + 2HCO_3^- + (1/4)CO_2 + (11/4)H_2O$$

Some of the released HCO_3^-, an alkalinity, may diffuse to the water surface and be lost by lateral water movement. During the next drying phase, the FeS precipitated in the previous submergence phase is reoxidized to produce protons together with SO_4^{2-}:

$$FeS + (9/4)\,O_2 + (5/2)\,H_2O = Fe\,(OH)_3 + 2H^+ + SO_4^{2-}$$

Thus, the soil will be acidified.

Conversely, when a soil has a pH below 4.5 and a small amount of SO_4^{2-}, either as an adsorbed anion or basic aluminum sulfate [$Al(OH)SO_4$], SO_4^{2-} remains unchanged because sulfate reducers are inactive at pH values below 5. Thus the concentration of SO_4^{2-} in the soil solution increases gradually during the submergence phase:

$$ads.\ SO_4^{2-} + 2H_2O = ads.\ (OH)_2 + 2H^+ + SO_4^{2-}$$
$$Al\,(OH)\,SO_4 + 2H_2O = Al(OH)_3 + 2H^+ + SO_4^{2-}$$

The protons formed in these processes will be used for the reduction of ferric iron, taking over this role from CO_2 under conditions too acidic to effect appreciable dissociation of aqueous CO_2 (pH < 6.4):

$$Fe(OH)_3 + (1/4)CH_2O + 2H^+ = Fe^{2+} + (1/4)CO_2 + (11/4)H_2O$$

In the soil solution, apparent dissolution of $FeSO_4$ is observed. Some of the $FeSO_4$ will diffuse into the soil-water interface and undergo the following reaction:

$$Fe^{2+} + SO_4^{2-} + (1/4) O_2 + (5/2) H_2O = Fe(OH)_3 + 2H^+ + SO_4^{2-}$$

The iron is precipitated as hydrated ferric oxides, while any sulfuric acid formed will be lost laterally with surface water movements. This is a partial deacidification. The formed sulfuric acid might also be lost through percolation when soil permeability is high, as in many Japanese paddy soils.

Once pH increases above 4.5–5 during reduction, weak acids such as aqueous CO_2 ($pK_1 = 6.4$) or H_2S ($pK_1 = 7.0$) begin to function as proton donors in iron reduction. This implies that at a low pH in the presence of exchangeable SO_4^{2-} or basic sulfates, acidity will be lost until a pH of 4.5–5 is reached.

Acid sulfate soils deacidify in the above mechanism to a pH of around 5, but if the soil pH rises above 5 an opposite acidification mechanism operates to keep the soil pH around 4.5–5. Conversely, if oxidation of FeS formed during submergence of a non-acid marine soil leads to pH values well below 5 in the subsequent dry period, some of the acid formed will be removed according to the deacidification mechanism during the subsequent submergence period.

Thus, alternating reduction and oxidation always proceed in the direction of zero mineral acidity or zero carbonate alkalinity; that is, towards equal concentrations of H^+ and HCO_3^-. In such a situation, the pH attains the following value[1]:

$$pH = (pK_H + pK_1 - \log P_{CO2})/2$$

where K_H is the Henry's law constant, K_1 the first dissociation constant of carbonic acid, and P_{CO2} the partial pressure of CO_2. At 25°C the relationship is given as follows:

$$pH = 3.95 - \log P_{CO2} / 2$$

At P_{CO2} of 10^{-2} atm, pH is 4.9, within the empirically known range of soil pH in deltas and polder lands.

7.4 Morphological changes

As discussed in Chapter 3, ferric and manganic compounds are reduced to more highly mobile ferrous and manganous ions in submerged soils. Where there is a moderate downward movement of water through the solum, the ferrous and manganous ions are leached from the surface plow layer into the relatively more oxidized environment of the subsurface and subsoil, and are then precipitated to form illuvial horizons. As the process is governed by their respective redox potentials, manganic and ferric oxides tend to precipitate separately, forming an iron illuvial horizon that is underlain by a manganese illuvial horizon. There may be two explanations for this:

- The deeper the layer in the solum, which is further away from the plow layer, the higher the redox potential, and manganic ions precipitate in a more oxidized environment than ferrous ions;
- The standard redox potential is higher for the manganic-manganous system, so manganous ions move down first and are precipitated where the redox potential is high enough to allow precipitation. Later, iron, which is more difficult to reduce, begins to be mobilized. When it is leached down to the zone of manganese precipitation, it reduces manganese spontaneously. Thus, the iron is oxidized and precipitated, and manganese is displaced further downward.

A typical morphology of this kind, as illustrated in Figure 7.2 (Mitsuchi, 1981), has been reported in Japan and elsewhere, notably in the *sawah* soils of West Java, Indonesia (Koenigs, 1950). Kyuma and Kawaguchi (1966) once called paddy soils with this morphology 'aquorizems' to distinguish them from other paddy soils. The process leading to aquorizem morphology may be termed 'aquorization', the concept of which was developed from the studies of Wada and Matsumoto (1973) and Mitsuchi (1974a, 1975). Mitsuchi revealed that the aquorizem profile features are modified by the permeability of the profile, which is in turn regulated by texture and pore size distribution.

The morphological features can be examined in more detail, using Figure 7.2. Both soils in the figure may be classified in their original state as brown lowland soil (Japanese system) or Dystric Eutrudept (Soil Taxonomy). They occur on natural levees and have a loam to a fine sandy loam texture. Before they were turned into paddy fields, these soils were highly

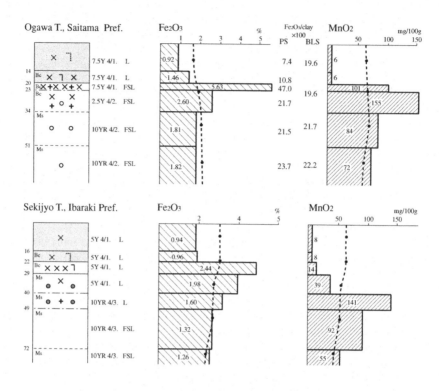

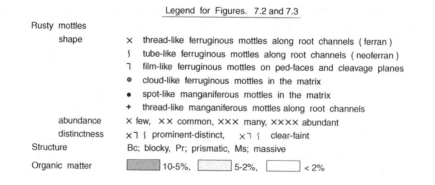

Figure 7.2
Highly permeable lowland paddy soils derived from brown lowland soils (Fluvents).
(dotted line: adjacent non-submerged soils) (Source: Mitsuchi, 1981)

permeable. However, puddling caused the formation of a traffic pan below the plow layer, which lowered permeability and helped to retain water for submergence. Thus, the redox condition within the profile is quite contrasting—strongly anaerobic in the plow layer during submergence, but

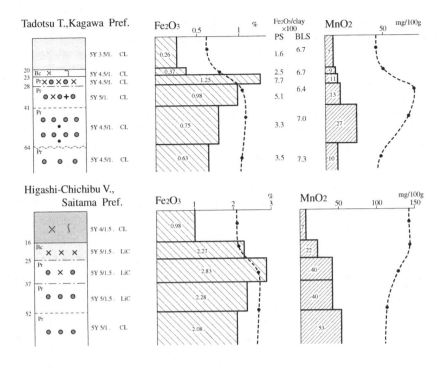

Figure 7.3
Slowly permeable lowland paddy soils derived from brown lowland soils (Fluvents).
(dotted line: adjacent non-submerged soils) (Source: Mitsuchi, 1981)

quite aerobic in the subsoil throughout the year. The gray matrix color accompanied by segregation of iron into rusty mottles in the upper part of the profile is due to surface gleying (or inverted gleying), while the original non-reduced brown color is retained in the subsoil. The iron and manganese distribution patterns in the profile also clearly reflect the contrasting redox condition. Compared with the distribution of iron in the adjacent soil, that was used for upland crops (dotted lines), the paddy soil lost iron from the plow layer and the plow pan layer, but gained a roughly corresponding amount of iron in the subsurface horizons. A similar change can be seen in the distribution of manganese, although there is a slight downward shift compared with iron, as predicted from the redox behavior of these two elements.

A similar morphological feature is observed in paddy soils with a finer texture and slow permeability. This variant is schematically shown in Figure 7.3 (Mitsuchi, 1981). The soil is a brown lowland soil or Typic Udifluvent that, in its original state, has no mottling, at least to a depth of 1 m. This soil can also occur on a natural levee, but its texture is finer, clay loam, and soil material is packed more densely throughout the profile, lowering its permeability. The gray color in the surface horizons, with thread-like and filmy mottles, results from surface gleying under submergence. Even in this slowly permeable profile, an eluviation-illuviation process involving iron, and to some extent, manganese is evident from their vertical distribution. In this kind of soil, iron accumulation occurs commonly in the transitional zone between the gleyed surface horizon and the pseudogleyed subsoil horizon (see below). Ferrous and manganous ions carried down by percolating water are first adsorbed by the soil via cation exchange, and are later deposited as oxides upon aeration following the drainage of surface water (Wada and Matsumoto, 1973). As the reduction is much less intense in the subsurface horizon, most of the mobilized iron oxides remain there and the iron content gradually increases. However, the accumulation of iron and manganese is only relative, and the simultaneous dissolution and removal of iron and manganese from the subsoil by reducing substances in the percolating water partially offsets their accumulation by illuviation.

However, the gray-colored subsoil horizons resemble those of pseudo-gley soils (Typic Epiaquept, Aeric Epiaqualf or Epiaquult), or have low chroma mottles and streaks spreading along the voids (ped faces and channels) and high chroma rusty mottles within the matrix of the peds. Thin section observation revealed that there was no difference in grain-size distribution between the gray colored ped faces and the matrix, suggesting that the gray colored ped faces are neither clay skins nor flood coatings (or gleyans). The pseudogleyed subsoil seemed to have formed partly as a result of the dissolution of iron on ped faces and channel walls by reducing organic substances percolating through the voids (Wada et al., 1971 a,b; Okazaki and Wada, 1976), and partly as a result of in situ segregation of iron in the matrix. Such an irrigation-induced pseudogley feature is another important morphological alteration and is called 'grayization' or 'gray coloring' in Japan (Kyuma et al., 1987).

This irrigation-induced pseudogleying or grayization is, by itself (even unaccompanied by the oxidative illuviation of iron and manganese), an important change in the sense that a non-aquic soil assumes the morphology of an aquic soil to a considerable depth under artificial submergence. If this change is not strongly developed and the boundary between the grayized horizon and an underlying oxidized horizon appears within 1 m of the soil surface, an irrigation-induced aquic or anthraquic moisture regime (see below) may be recognized. However, if the grayization is strongly developed and irrigation-induced pseudogleying goes deeper, it is difficult to distinguish it from natural pseudogleys. Grayization takes place in paddy soils on relatively elevated parts of recent flood plains (such as river levees), while natural pseudogleys appear to be restricted to older land surfaces. This difference in soil age produces differences in; the content and distribution of organic carbon in the subsoil, color contrast of the subsoil mottles, the degree of compaction, and the degree of crystallization of iron oxides or, more generally, the degree of weathering of the subsoil.

Grayization or gray coloring is related clearly to the ferrous iron content of the soil. Compared with upland soils derived from the same parent material, paddy soils tend to show low chroma or no chroma and gray color. As shown in Figure 3.8, quite a large amount of Fe^{II} remains in paddy fields, even after they are drained (Motomura, 1969). Mitsuchi (1974a) also studied the amount of different forms of Fe^{II} remaining after drying of paddy soils (Table 7.1). Motomura and Yamanaka (1963) studied the relationships between the field matrix color of the soil and the Fe^{II} content, and found the following:

$$10YR = 4.4 \text{ mg g}^{-1} \text{ soil} \qquad 2.5GY = 15.7 \text{ mg g}^{-1} \text{ soil}$$
$$2.5Y = 8.2 \qquad\qquad\qquad 5GY = 18.0$$
$$5Y = 7.5 \qquad\qquad\qquad 7.5GY = 22.2$$
$$7.5Y = 11.4 \qquad\qquad\qquad 10GY = 17.3$$
$$10Y = 15.6$$

Table 7.1 Changes in Fe^{II} and soil color in submergence and drainage treatment

Soil sample (location and horizon)	Soil type	Reduced soil			Air-dried soil		
		$\dfrac{\text{T-Fe}^{II}}{\text{Free Fe}}$ %	$\dfrac{\text{A-Fe}^{II}}{\text{T-Fe}^{II}}$ %	Soil color	$\dfrac{\text{R-Fe}^{II}}{\text{Free Fe}}$ %	$\dfrac{\text{R-Fe}^{II}}{\text{T-Fe}^{II}}$ %	Soil color
Nagano Exp. Sta., C	Lowland soil	78.8	58.4	2.5GY3.8/1.5	27.0	34.2	4.2Y4.0/2.0
Higashi-Chichibu, C	Lowland soil	56.8	54.4	5.0Y3.8/2.0	25.2	44.3	2.0Y4.0/2.3
Hata, Nagano, B	Kuroboku (Ando)	38.8	62.1	2Y3.7/2.8	5.7	17.4	9YR4.1/3.7
Nishigahara, Subsoil	Kuroboku (Ando)	24.7	92.5	1.5Y3.7/2.8	3.5	14.4	9YR4.1/3.6
Mikatagahara, B	Red-yellow soil	14.7	57.4	4Y4.0/3.4	2.8	19.7	9.5YR4.6/5.1

Notes:

Free Fe: Free iron oxides determined by Mg-ribbon reduction method

T-Fe^{II}: Difference in total Fe^{II} determined by HF-H$_2$SO$_4$ digestion method before and after submergence

A-Fe^{II}: Active Fe^{II} determined in N-NaOAc extract at pH 2.8

R- Fe^{II}: Residual Fe^{II} determined as total Fe^{II} after air-drying

Reduced soil: Submerged for 2 months at 28°C with 0.5% rice straw added

Air-dried soil: Air-drying the reduced soil for 2 weeks and then pulverized in a mortar

(Source: Mitsuchi, 1974a)

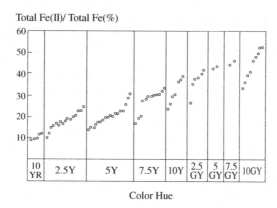

Figure 7.4
Relationship between color hue and the ratio of total Fe[II] to total Fe (Source:
Motomura and Yamanaka, 1969)

The correlation was even better when the ratio of Fe[II] to total Fe was used, as in Figure 7.4. This suggests that the gray color is a result of the mixing of hydrated ferric oxides that have reddish-yellow hues and high chroma with Fe[II] that has bluish-green hues and low chroma.

A leaching loss of iron from the surface plow layer of paddy soils as a result of percolation under the reduced condition may explain some of the gray coloring. However, Mitsuchi (1974a) claimed that the change in the valence state of iron is the main cause of the gray coloring of the plow layer of paddy soils.

7.5 Morphological features relevant to the classification of paddy soils

Some of the issues relevant to the classification of paddy soils should be considered, in relation to the morphological changes that are induced by the long-term utilization of soil as a paddy field.

It is obvious that originally aquic soils would not show any clearly visible morphological changes, even after long-term use as a paddy field. Because these soils normally have a shallow ground-water table, the whole profile is water saturated and reduced during the submerged period of rice cultivation, and no downward movement of water occurs. In contrast, soils that maintain a ground-water table that is sufficiently low, even under submergence, to allow water to percolate through the profile, develop illuvial horizons of iron and manganese in the subsurface layers.

In order to distinguish these moisture regimes, Soil Survey Staff (1999) define three types of water saturation:

- Endosaturation—The soil is saturated with water in all layers from the upper boundary of saturation to a depth of 200 cm or more from the mineral soil surface.
- Episaturation—The soil is saturated with water in one or more layers within 200 cm of the mineral soil surface, and has one or more unsaturated layers, with an upper boundary above a depth of 200 cm, below the saturated layer. The zone of saturation, i.e., the water table, immediately above a relatively impermeable layer.
- Anthric saturation—This term refers to a special kind of aquic condition that occurs in soils that are cultivated and irrigated (flood irrigation). Soils with anthraquic conditions must meet the requirements for aquic conditions and have both of the following characteristics:
    (1) a tilled surface layer and a directly underlying slowly permeable layer that has, for three months or more in normal years, both
      (a) Saturation and reduction; and
      (b) Chroma of 2 or less in the matrix; and
    (2) A subsurface horizon with one or more of the following characteristics:
      (a) Redox depletions with a color value, moist, of 4 or more and chroma of 2 or less in macropores; or
      (b) redox concentrations of iron; or
      (c) two times or more the amount of iron (by dithionite citrate) contained in the tilled surface layer.

'Soils with anthraquic conditions' that are described under anthric saturation, appear to encompass most artificially irrigated paddy soils. Therefore, what are anthraquic conditions? They are a variant of episaturation and are associated with controlled flooding (for crops such as wetland rice and cranberries), which causes reduction processes in the saturated, puddled surface soil and oxidation of reduced and mobilized iron and manganese in the unsaturated subsoil.

Redox concentrations and depletions, collectively called 'redoximorphic features', in the above definition of soils with anthraquic conditions, are further defined as follows:

- Redox concentrations—These are zones of apparent accumulation of Fe-Mn oxides, including: (1) Nodules and concretions, which are cemented bodies that can be removed from the soil intact. Concretions are distinguished from nodules on the basis of internal organization. A concretion typically has concentric layers that are visible to the naked eye. Nodules do not have a visibly organized internal structure. Boundaries commonly are diffused if formed *in situ* and sharp after pedoturbation. Sharp boundaries may be relict features in some soils; (2) Masses, which are non-cemented concentrations of substances within the soil matrix; and (3) Pore linings, i.e., zones of accumulation along pores that may be either coatings on pore surfaces or impregnations from the matrix adjacent to the pores.

- Redox depletions—These are zones of low chroma (chromas less than those in the matrix) where either Fe-Mn oxides alone or both Fe-Mn oxides and clay have been stripped out, including: (1) Iron depletions, i.e., zones that contain low amounts of Fe and Mn oxides, but have a clay content similar to that of the adjacent matrix (often referred to as 'albans' or 'neoalbans'); and (2) Clay depletions, i.e., zones that contain low amounts of Fe, Mn, and clay (often referred to as silt coatings or skeletons).
- Reduced matrix—This is a soil matrix that has a low chroma *in situ*, but undergoes a change in hue or chroma within 30 minutes after the soil material has been exposed to air.
- In soils that have no visible redoximorphic features, a reaction to an α, α'-dipyridyl solution satisfies the requirement for redoximorphic features.

In the revised Soil Taxonomy (Soil Survey Staff, 1999), only one anthraquic subgroup is set up; that is the 'Anthraquic Eutrudept'. However, anthraquic subgroups of many other great groups could and should be recognized, such as Anthraquic Dystrudept, Anthraquic Hapludult, Anthraquic Hapludalf, and so forth.

Paddy soils occurring on originally aquic lands are mostly classified within the framework of the aquic suborders of various orders. As stated in Chapter 1, paddy soils can be developed in all of the orders except for Gelisols.

Notes

(1) When there is no alkalinity, or mineral acidity, water in which carbon dioxide is dissolved will have the following charge balance:

$$[H^+] = [HCO_3^-] + 2[CO_3^{2-}] + [OH^-]$$

As water retains slightly acidic pH, $[H^+]$ should be balanced by $[HCO_3^-]$, or $[H^+] = [HCO_3^-]$. From the first dissociation constant of carbonic acid, the following equation is derived:

$$K_1 = [HCO_3^-][H^+] / [H_2CO_3]$$

$$\text{As } [H_2CO_3] = K_H \cdot P_{CO2}$$

$$[H^+]^2 = K_1 \cdot K_H \cdot P_{CO2}$$

$$\text{Thus: } pH = (pK_H + pK_1 - \log P_{CO2}) / 2$$

References

Brinkman, R. 1970. Ferrolysis, a hydromorphic soil forming process. *Geoderma*, 3: 199–206.

Brinkman, R. 1977. Surface-water gley soils in Bangladesh: Genesis. *Geoderma*, 17: 111–144.

Brinkman, R. 1978. Ferrolysis: chemical and mineralogical aspects of soil formation in seasonally wet acid soils, and some practical implications. In IRRI (Ed.) *Soils and Rice*, pp. 295–303. IRRI, Los Baños, Philippines.

Kanazawa, S. and Yoneyama, T. 1980. Microbial degradation of ^{15}N-labeled rice residues in soil during two years' incubation under flooded and upland conditions. *Soil Sci. Plant Nutr.*, 26: 229–239.

Kawaguchi, K. and Kawachi, T. 1969 a. Cation exchange reactions in the submerged soils. *J. Sci. Soil & Manure, Japan*, 40: 89–95. (In Japanese).

Kawaguchi, K. and Kawachi, T. 1969 b. Changes in the composition of exchangeable cations in the course of drying of preliminarily submerged soils. *J. Sci. Soil & Manure, Japan*, 40: 177–183 (In Japanese).

Koenigs, F.F.R. 1950. A 'sawah' profile near Bogor (Java). *Trans. 4th Int'l Congr. Soil Sci.*, 1: 297–300.

Kubota, S. 1961. *Characteristics of polder land soils and the transition of the soil types after poldering*. Spec. Rep. Okayama Agric. Expt. Sta.No.59. (In Japanese).

Kyuma, K. 1978. Mineral composition of rice soils. In IRRI (Ed.) *Soils and Rice*, pp. 219–235. IRRI, Los Baños, Philippines.

Kyuma, K. and Kawaguchi, K. 1966. Major soils in Southeast Asia and the classification of soils under rice cultivation. *Tonan Ajia Kenkyu (Southeast Asian Studies)*, 4: 290–312.

Kyuma, K., Mitsuchi, M. and Moormann, F.R. 1987. Man-induced soil wetness: the "anthraquic" soil moisture regime. *Proc. 9th Int. Soil*

Classification Workshop: Properties, Classification and Utilization of Andisols and Paddy Soils., pp. 138–146.

Mitsuchi, M. 1974 a. Genetic characteristics of lowland paddy soils and their implications on the soil classification. *Bull. Nat. Inst. Agric. Sci.*, B25: 29 –115. (In Japanese).

Mitsuchi, M. 1974 b. Chloritization in lowland paddy soils. *Soil Sci. Plant Nutr.*, 20: 107–116.

Mitsuchi, M. 1974 c. Characters of humus formed under rice cultivation. *Soil Sci. Plant Nutr.*, 20: 249–259.

Mitsuchi, M. 1975. Characters of upland paddy soil on the Mikatagahara terrace, Shizuoka Prefecture. *J. Sci. Soil & Manure, Japan*, 46: 333–339. (In Japanese).

Mitsuchi, M. 1981. Characteristic features of paddy soils of Japan. *Proc. Symp. on Paddy Soils, Nanjing*, pp. 419–427.

Motomura, S. 1969. Behavior of ferrous iron and its roles in paddy soils. *Bull.Nat. Inst. Agric. Sci.*, B21: 1–114. (In Japanese).

Motomura, S. and Yamanaka, K. 1963. Studies on the gley formation of soils (Part 4). Development of reduced soil color in relation to ferrous iron. *J. Sci. Soil & Manure, Japan*, 34: 428–432. (In Japanese).

Okazaki, M. and Wada, H. 1976. Some aspects of pedogenic processes in paddy soils. *Pedologist,* 20: 139–150.

Soil Survey Staff 1999. *Soil Taxonomy, 2ⁿᵈ Ed.: A Basic System of Soil Classification for Making and Interpreting Soil Surveys.* USDA Agric. Handb. No.436, U.S. Gov't. Printing Office, Washington, D.C.

Van Breemen, N. 1975. Acidification and deacidification of coastal plain soils as a result of periodic flooding. *Soil Sci. Soc. Amer. Proc.*, 39: 1153–1157.

Wada, H. and Matsumoto, S. 1973. Pedogenic processes in paddy soils. *Pedologist*, 17: 2–15.

Wada, H., Yoshida, H. and Takai, Y. 1971 a. Cutans developed in subsurface horizons of clayey paddy soils (Part 1). *J. Sci. Soil & Manure, Japan*, 42: 12–17. (In Japanese).

Wada, H., Yoshida, H. and Matsumoto, S. 1971 b. Cutans developed in subsurface horizons of clayey paddy soils (Part 2). *J. Sci. Soil & Manure, Japan*, 42: 65–68. (In Japanese).

Yamazaki, S. and Saeki, T. 1980. Weeds in paddy soils and the light environment. In Takai, Y. (Ed.) *Report of Researches on Environmental Sciences, B18-R.-12-1*, pp. 8–17. (Cited from Kanazawa, S. 1982. Chapter 6. Biota in paddy soils. In Yamane, I. (Ed.) *Paddy Soil Science*, pp. 233–279. No-Bun-Kyo, Tokyo. In Japanese).

Yoshida, M. and Itoh, N. 1974. Composition of exchangeable cations in submerged paddy soil and acidifying process associated with exposure to air. *J. Sci. Soil & Manure, Japan*, 45: 525–528. (in Japanese).

Chapter 8: Fertility Considerations for Paddy Soils (I)—General Nutrient Balance and Nitrogen

8.1 Nutrient balance in rice cultivation

The nutrient balance in rice cultivation is discussed here using examples from Japan. Table 8.1 presents the major elements needed to grow one crop of paddy rice, including fertilizer input. Although the yield of brown rice in the table is a little higher than the national average of approximately 5 t ha^{-1}, the figures are the means of thirteen locations throughout Japan, and are considered to represent the current nutrient situation of Japanese rice-farming. Generally the rate of N application has decreased recently from about 100 kg to 80 kg, mainly due to the environmental concerns of farmers. The table shows that with this rate of application, it is essential to return crop residues to the soil to maintain nitrogen levels. This also

Table 8.1 NPK balance sheet and NPK required to produce one ton of brown rice[1]

Yield of brown rice (t ha^{-1})	Input/output and balance	N (kg ha^{-1})	P$_2$O$_5$ (kg ha^{-1})	K$_2$O (kg ha^{-1})
	Nutrients applied (A)	80	74	81
	Uptake of nutrients by grains and straw (B)	118	61	159
5.79	Uptake of nutrients by grains (C)	74	43	30
	Balance (A–B)[2]	–38	13	–78
	Balance (A–C)[2]	6	31	51
Nutrients required to produce 1 t of brown rice (kg)[3]		20.4	10.5	27.5
Nutrients required to produce 6 t of brown rice (kg)		122	63	165

Notes:
1) Averages of data at 13 locations from northern through southern Japan.
2) Differences between input and uptake of nutrients. Negative signs denote deficit situation in the soil.
3) Amount of nutrients taken up divided by yield.
(Source: Suzuki, 1997)

Table 8.2 Nutrient balance in rice cultivation (in kg ha⁻¹)—A case study

Input				Output			
Source	N	P_2O_5	K_2O	Source	N	P_2O_5	K_2O
Mineral	95.3	103.0	89.0	Uptake by rice	95.8	50.2	158.0
Organic	20.1	6.0	16.1				
N-fixation	40	-	-	Denitrification	70	-	-
Irrigation	16.5	0.8	33.0	Leaching	20.0	2.1	26.7
Rain	5.0	0.5	5.3	Runoff	0.6	0.0	0.9
Total	176.9	110.3	143.4	Total	186.4	52.3	185.6

(Source: Asano and Yatazawa, 1976)

applies to potassium. Only phosphorus appears to be sufficient, or rather it tends to be accumulated in the soil. However, this table does not illustrate the fate of each nutrient element.

Table 8.2 is another nutrient budget taken for rice cultivation. As it considers some other inputs and outputs, it provides more detail on the fate of the nutrients. There is substantial N fixation in this case, but a higher N loss through denitrification. This particular field is discharging more nutrients in drained water than those it receives from irrigation. Again, phosphorus tends to be accumulated due to fertilization of the soil, but as a whole, N and K_2O appear to be deficient.

In order to understand the real nutrient status, it is necessary to examine the nutrient dynamics of the soil. The following section does this.

8.2 Nitrogen balance

8.2.1 Nitrogen balance in paddy soils

N is by far the most important mineral nutrient of rice, as seen in Table 8.3. This table is very revealing, and here the no-N plot is compared with the no-fertilizer plot. The yield indices for both plots are similar showing that N is vital for rice production.

Hasegawa (1992) studied, the N balance of paddy soil near Lake Biwa (Shiga Prefecture, Japan) by differentiating soil and fertilizer N, as in Figure 8.1. From this figure, the locations of fertilizer N and soil N can be clarified, as shown in Table 8.4.

It is interesting to note the following from Table 8.4:
- Of the total uptake of N by rice, 60% comes from soil N and 40% from fertilizer N, confirming what is generally said about the N nutrition of rice.

Table 8.3   Results of fertilizer trials in Japan in terms of the mean yield indices of variously treated plots relative to the complete plot yield

Crop	Trial type	Treatment					Number of trials
		Complete	No-N	No-P$_2$O$_5$	No-K$_2$O	None	
Rice	Pot	100	55	89	88	53	2850–2898
	Field	100	83	95	96	78	1161–1187
Upland rice	Pot	100	49	79	92	41	176
	Field	100	51	84	75	38	117–126
Upland cereals*	Pot	100	34	62	81	26	1937–1991
	Field	100	50	69	78	39	822–841

* Such crops as barley, naked barley and wheat are included
(Source: Kawasaki, 1953)

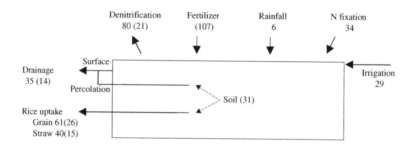

Figure 8.1
N balance in a paddy soil (in kg N ha^{-1}) (Figures in parentheses: fertilizer N)
(Source: Hasegawa, 1992)

- The overall deficit of soil N, even after considering the fertilizer N remaining in the soil, is compensated by the mineralization of soil organic N; which can be replenished by N fixation and/or returning the harvested straw.

Much more research on the nitrogen budget of rice growing needs to be conducted in different places for different methods of rice cultivation.

The major pathways of the inputs and outputs of N in paddy soils are given in the N balance studies above. The following section examines each of these pathways in some detail. Also, important pathways found in other countries, such as input through the absorption of atmospheric NH$_3$ and output through NH$_3$ volatilization, is also considered.

Table 8.4   Nitrogen balance in a paddy soil in Shiga Prefecture, Japan

	Fertilizer N			Soil N		
Balance	Item	kg ha⁻¹	%	Item	kg ha⁻¹	Sum
Input	Fertilizer applied	107	100	N fixation	34	69
				Irrigation	29	
				Rainfall	6	
Output	Rice uptake	41	38	Rice uptake	60	140
	grain	26	24	grain	35	
	straw	15	14	straw	25	
	Denitrification	21	20	Denitrification	59	
	Drainage	14	13	Drainage	21	
	primary*	6	6			
	secondary	8	7			
Difference	Remaining in the soil	31	29	Fertilizer	31	31

* Primary load means the part lost in drained water right after application, whereas secondary load means the part once immobilized and then lost in drained water.
(Source: Hasegawa, 1992)

8.2.2 Addition of nitrogen by irrigation water

In Japan, a concentration of N in water that is higher than 0.2 mg litre⁻¹ is considered to be an N eutrophication hazard because unpolluted water normally contains N at concentrations of below 0.2 mg litre⁻¹. If irrigation water used for a single crop of rice is assumed to be 1000 mm, 1 mg litre⁻¹ of an element in the irrigation water would supply about 10 kg of that element per hectare per crop. Therefore, the supply of N by unpolluted irrigation water does not normally exceed 2–3 kg ha⁻¹, a small figure relative to the total N uptake of rice plants from sources other than fertilizers (see Table 8.5).

However as shown by nutrient balance studies, the contribution of irrigation water to the N status of rice is often quite substantial, ranging somewhere from 15 to 30 kg ha⁻¹. In fact, Kobayashi (1971) gave a figure of 1.15 mg litre⁻¹ as the mean total N concentration of Japanese river water, which may supply 15 to 20 kg ha⁻¹ of N to a crop of rice with some 1500 mm of irrigation water. Takamura and Tabuchi (1977) gave a figure of 27.8 kg N ha⁻¹ as the national average of N from rainfall and irrigation water, meaning some 20 kg ha⁻¹ or more of N comes from irrigation water.

8.2.3 Addition of nitrogen by rain water

In the atmosphere, some N is naturally fixed as NO_x by high energy processes such as lightning and cosmic ray irradiation. This source of N is said to

Table 8.5 Nitrogen uptake by wetland rice from sources other than fertilizer

Region	N absorbed by crop, kg/ha	Method	Authors
Hokkaido-Kyushu, Japan	64 (47–113)[1]	Non-N plot[2]	Yanagisawa and Takahashi (1964)
Fukui and other regions, Japan	30–100	Isotope[3]	Koyama (1975)
86 exp't plots throughout Japan	66(29–99)	Non-N plot	Toriyama (2002)
Bang Khen, Thailand	37	Non-N plot	Koyama *et al.* (1975)
	52	Isotope	Koyama *et al.* (1975)
Los Baños, Philippines	45–80	Non-N plot	Shiga and Ventura (1976)
	97–113[4]	Isotope	IRRI (1974)
Muara, Indonesia	47	Non-N plot	Isumanadji *et al.* (1973)
Pusakanegara, Indonesia	65	Non-N plot	Isumanadji *et al.* (1973)

Notes: 1) Average (maximum-minimum) of 15 provincial agricultural experiment stations. 2) N-absorption in non-N fertilizer plots. 3) Estimated by N^{15}-labeled N fertilizer. 4) Dry season, 1973.
(Source: Watanabe, 1978; Toriyama, 2002)

account for only a few kg ha^{-1} y^{-1} to at most 10 kg ha^{-1} y^{-1}, but probably 5–6 kg ha^{-1} per cropping season (as seen in Tables 8.2 and 8.4) would be a reasonable estimate. However nowadays, a much higher amount of N as NO_3^--N and NH_4-N is precipitated locally in wet, as well as dry acid depositions. One report from the Netherlands estimated that about 64 kg N $ha^{-1}y^{-1}$ comes from dry and wet depositions (van Breemen *et al.*, 1982), mainly as a result of the volatilization of NH_3 from large-scale intensive animal husbandry.

8.2.4 Absorption of ammonia from the atmosphere

It has been confirmed that ammonium ions in the atmosphere are readily absorbed by plants and soil (Denmead *et al.*, 1976; Hutchinson *et al.*,

Table 8.6 Comparison of amounts of biologically fixed N in different ecosystems

Ecosystem	Fixed N in kg ha^{-1}
Upland field	7–28
Paddy field	13–99
Grassland	
Nonleguminous	7–114
Legumes + grass	73–865
Forest	58–594

(Source: Hauck, 1971)

1972). Hanawall (1969) presented data showing that soil had the capacity to absorb as much as 55–74 kg N ha^{-1} per year, when the atmospheric ammonia concentration is 38g N m^{-3}. A study by the International Rice Research Instutute (IRRI), showed that the atmosphere above a non-fertilized IRRI rice field contained 8–35 g N m^{-3} of ammonia. Thus, this source of N can be of some significance locally.

8.3 Nitrogen fixation

8.3.1 Nitrogen enrichment by N fixation

Biological N fixation has been shown to be more significant in paddy soils than in ordinary upland soils. Hauck (1971) provided the data in Table 8.6, which clearly shows the general advantage of paddy soils over upland soils.

Assuming that 30 kg ha^{-1} of N$_2$ are fixed in a paddy field per year, Quispel (1974) calculated the total amount of N fixed in the whole world's paddy fields to be 4 million tons per year, an amount equivalent to more than 11% of the fertilizer N applied annually to all the crops in the world at that time.

In Japan, Ono and Koga (1984) estimated the fixation of N by paddy soils to be 25–35 kg N ha^{-1} y^{-1}, using an analysis of enriched N in the surface soil. Ito (1977) analyzed the N balance in a long-term fertilizer trial for rice of over 70 years and found that the annual average N enrichment in the surface soil was 38.5 kg N ha^{-1} for the non-fertilized plot and 39.6 kg N ha^{-1} for a plot with liming.

8.3.2 Nitrogen fixing microorganisms and their function

The importance of biological N fixation in paddy fields was first reported on blue-green algae (De, 1939). The list of N fixing microorganisms has expanded profusely[1] as techniques to study N fixation have been developed.

There are three methods to determine the rate of N fixation:
- measuring the total N increase by the Kjeldahl method
- the ^{15}N isotope dilution method
- measuring the acetylene reduction activity.

The relative sensitivity of these methods has increased respectively from 1 via 100 to 100,000. The last method can be used in the field, so is now almost exclusively used for determining N-fixation.

The method is based on the fact that nitrogenase (N fixing enzyme) reduces acetylene, C$_2$H$_2$, to ethylene, C$_2$H$_4$, in the same process that reduces N$_2$ to NH$_3$, as follows:

$$N_2 + 6H^+ + 6e + 12ATP \rightarrow 2NH_3 + 12ADP + 12P_i$$
$$3C_2H_2 + 6H^+ + 6e + 12ATP \rightarrow 3C_2H_4 + 12ADP + 12P_i$$

where P_i is inorganic phosphorus. As the nitrogenase has a high affinity for acetylene relative to N_2 (even in the presence of N_2), nitrogenase selectively reduces acetylene when a small amount of acetylene is added as the substrate.

According to the above reactions, a conversion factor of 3 is used to calculate the fixed amount of N_2 from the measurement of acetylene reduction activity. However, it is known that the conversion factor may vary from 3 to more than 4, depending on the conditions (such as the availability of energy sources). This makes an exact quantification of N fixation difficult.

More recently, the above reactions are often written by considering the ATP dependent production of H_2:

$$N_2 + 8H^+ + 8e + 16ATP \rightarrow 2NH_3 + 16ADP + 16P_i$$
$$4C_2H_2 + 8H^+ + 8e + 16ATP \rightarrow 4C_2H_4 + 16ADP + 16P_i$$

In this case, the theoretical conversion factor is 4.

Yamamuro (1987) introduced the $^{15}N_2$ gas feeding method to study N fixation in paddy fields. This method applies water containing dissolved $^{15}N_2$-labelled air to the surface water of a paddy field. It requires a special technique to analyze the $^{15}N_2$ atom percent excess, but it has definite advantages in quantitative analysis and in tracing metabolic N cycling.

The biochemical process of N fixation is represented schematically by Figure 8.2. Nitrogenase consists of Fe protein and MoFe protein, both being sulfide-containing proteins sensitive to O_2. According to Eady et al. (1972), nitrogenase loses 70% of its activity when exposed to the air, with P_{O2} of 0.2 atm, for a minute. The maximum N fixation takes place at $P_{O2} = 0$ atm for facultative and obligate anaerobes, at $P_{O2} = 0.04$ atm for Azotobacters, and at $P_{O2} = 0.1$ atm for most of blue-green algae.

The presence of mineral N greatly inhibits microbial N fixation. It has been shown that NH_4^+ is more severely inhibiting than NO_3^- or NO_2^- to N fixation by Azotobacter and blue-green algae. NH_4^+ is known to inhibit the synthesis of nitrogenase but not its activity. It is also known that in the soil medium, the addition of NH_4^+ inhibits N fixation, but a simultaneous addition of glucose seems to lessen the inhibitory effect. In paddy soil growing rice plants, the addition of 800 mg kg^{-1} of NH_4-N lowered the N fixing capacity to 1/10 that of the control plot (Yoshida et al., 1973).

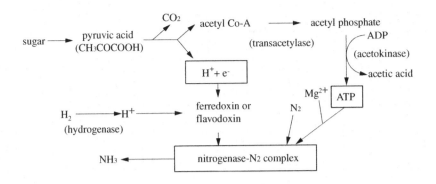

Figure 8.2
Schematic drawing of nitrogen fixation pathways (Source: Nakamura, 1970; Eady
& Postgate, 1974)

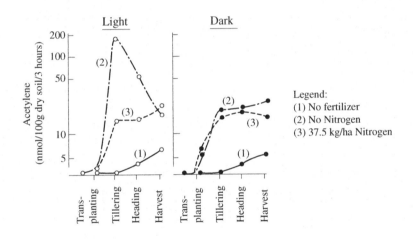

Figure 8.3
Influence of fertilizer application on acetylene reduction activity in a Thai soil
(0–2cm plow layer) (Source: Matsuguchi et al., 1978)

Matsuguchi *et al.* (1977) showed that for Thai soil, even 30 mg kg^{-1} of NH_4-N (37.5 kg NH_4-N ha^{-1}) decreased acetylene reduction remarkably and that the reduction was due to lowering of N fixation activity of photosynthetic microorganisms (see Figure 8.3).

In paddy soils under submergence, the reduction of soil and the supply of organic matter from weeds and algae appear to promote N fixation by

Table 8.7 Effect of submergence on N fixation in paddy soils

Plot	N fixed in one cropping season kg ha⁻¹
Without rice plants	
Upland plot	2.7
Waterlogged plot	42.5
With rice plants	
Upland plot	5.4
Waterlogged plot	79.8

(Source: Yoshida, 1971)

microorganisms. Yoshida (1971) showed the effect of submergence on the amount of N fixation, as in Table 8.7.

Watanabe summarized the data on N fixation, as in Table 8.8. The variation of the reported values is still wide, and this may partly reflect the actual variation depending on the soil properties, and partly be caused by the difficulties in making measurements.

Matsuguchi *et al.* (1974) surveyed N fixation at 36 sites in Thailand. They found that in more than half the cases, N fixation was only less than 5 kg N ha⁻¹, and was estimated to be about 20 kg ha⁻¹ at better sites. They ascribed these low figures to the acidity and low phosphorus supplying capacity of the soils. Their survey also revealed that *Clostridium*, blue-green algae, and non-sulfur purple bacteria are dominant among the N fixing flora in Thai paddy soils, as seen in Table 8.9. In contrast, relatively fewer Azotobacters and sulfur purple bacteria (*Chromatiaceae*) were present in the Thai paddy soils as compared with the Japanese paddy soils studied by Ishizawa *et al.* (1970). Another feature was a very low count of *Beijerinckia*, despite their wide occurrence in tropical soils.

Inoculation of paddy soils with blue-green algae or *Cyanobacteria* was attempted in India and Japan to enhance N fertility, and it was shown to increase rice yield, and soil N after the cropping in some instances where N was not applied.

Azolla is known to be an important N fixer in tropical paddy soils. *Anabaena azollae*, a species of blue-green algae, becomes symbiotic with an aquatic fern, Azolla (*A. pinnata, A. caroliniana, A. filiculvides*, etc.), to fix N. According to Becking (1976), the amount of N fixed by *Azolla-anabaena* ranged from 103 to 162 kg N ha⁻¹ y⁻¹. At sometime during the 1970s, Azolla was used in 40% of the rice fields in North Vietnam as an effective bio-fertilizer. In southern China too, it used to be raised widely as a green manure in rice fields. However in 1992, the IRRI Annual Report remarked that the use of *Azolla-anabaena* may not always be profitable compared with N fertilization, if the cost of the following tasks is taken into consideration:

- care for the inoculum;
- labor for the incorporation into the soil;
- phosphate fertilization.

The use of green manuring of *Azolla* and *Sesbania* may be profitable in rain-fed rice areas where no return from fertilizer application is assured due to unstable water supplies.

The importance of rhizosphere bacteria in N fixation was clearly demonstrated and intensively studied in the 1970s (Yoshida and Ancajas, 1971; Dommergues *et al.*, 1973; McRae, 1974). Yoshida (1971) found that rice rhizosphere show a high acetylene reduction activity in some Philippine soils. He estimated that about half of the fixed N (80 kg N ha^{-1}) was due to rhizosphere bacteria and half due to other N fixing microflora.

Watanabe *et al.* (1977) isolated N fixing microorganisms from rice roots, including *Beijerinckia, Enterobacter, Spirillum lipoferum*. N gas appears to be transmitted from the atmosphere via the air transmitting system of rice plants for N fixation in the rhizosphere (Yoshida and Broadbent, 1975). As it is known that nitrogenase is sensitive to oxygen, the aerobic-anaerobic interface on rice roots with adequate supplies of carbohydrates and N gas should be an almost ideal place for active N fixation.

According to Arihara (2000), Yanni *et al.* (1997) found a unique natural endophytic association between *Rhizobium leguminosarum* bv. *Trifolii* and rice roots in Egypt. The N supplied by a rotation of clover-rice, replaced 25–33% of the recommended application rate of fertilizer N to rice. They found that rhizobia nodulating on clover roots invaded rice roots and attained an internal population density of up to 1.1×10^6 endophytes per gram (fresh weight) of rice roots. They reported that inoculation with two endophytic strains (E11 and E12) of *R. leguminosarum* significantly increased the grain yield, harvest index and fertilizer N recovery rate of field-grown hybrid rice (Giza 175) (see Table 8.11). However, it is necessary to further confirm the effect of such an association between endophytic diazotrophs and rice roots on the N nutrition of rice. Ladha *et al.* (1993) also studied the natural associations of endophytic diazotrophs in rice roots under a rice-*Sesbania* rotation in the Philippines.

In a recent review by Yoneyama and Akao (1998), the role of associative N fixation in the rice rhizosphere was reevaluated and given less importance as the source of N nutrition of rice, because of competitiveness

Table 8.8 Estimated N fixation in paddy fields

Authors	Condition	N-fixing sites or agents	Method	N in kg ha⁻¹ (period)
Yoshida and Ancajas (1973)	IRRI field flooded+planted dry season	algae+soil (water and soil samples)	acetylene reduction	63 (17 wk)
	flooded+planted wet season		in vitro assay for 5 h every 3 wk	52 (17 wk) (3:1)*
Matsuguchi et al. (1976)	Thailand soil 42 sites	phototrophic + heterotrophic (water and soil samples)	acetylen reduction in vitro assay for 3 h under dark anaerobic, dark aerobic, light anaerobic and light aerobic	0.5–54 (yr) (69 av.) (3:1)*
Watanabe et al.(1977)	IRRI field every 3 wk	water + rhizosphere + surface soil (non-fertilized)	in situ acetylene reduction, 24h assay, every 3 wk	50 (335 d) (3:1)*
Reynaud and Roger (1976)	Senegal rice soil	algae (water and surface soil)	in situ acetylene reduction (15 or 30mins)	few kg cultivation cycle, exceptionally 10–30kg (3:1)*
Broughton et al (1976)	Malaysia rice soil flooded period +drained period	water+soil	in situ acetylene reduction, 24h assay	0.06–0.3 (d) (4:1)* 22 (av.0.18/ day for 120-d crop)
Hirano (1958)	Japan, Shikoku paddy field inoculated with algae	algae	total N increase of surface soil (difference between inoculated and non-inoculated)	26 (1 crop)

* Conversion factor in acetylene reduction activity measurement
(Source: Watanabe, 1978)

Table 8.9 N fixing microflora and the annual amount of N fixation in Thai paddy soils

Soil group	Location	N-fixing microflora Azoto-bacter	Beijer-inckia	Clostridium butyricum	Non-sulfur purple bacteria	Blue green algae	N-fixation* kg ha⁻¹y⁻¹ (mean)
Marine clay	water	10^1		10^2-10^3	10^1-10^2	10^1-10^2	1.3–17.4
	soil	10^0-10^3		10^4-10^5	10^3-10^4	10^1-10^4	(9.3)
Brackish clay	water	$<10^0$		10^2-10^3	10^1	$<10^1$	2.7–4.3
	soil	$\leq10^1$		10^4-10^5	10^2-10^3	10^2	(3.4)
Freshwater alluv. soil	water	$<10^2$		10^2-10^4	10^1-10^3	10^1-10^3	4.2–53.9
	soil	10^1-10^4	$\leq10^1$	10^4-10^6	10^3-10^5	10^3-10^4	(17.5)
Low humic gley soil (1)	water	$<10^0$		10^2-10^3	10^1-10^2	$<10^1$	0.5–3.9
	soil	$<10^1$	$<10^1$	10^3-10^4	10^3-10^4	10^2-10^3	(2.1)
Low humic gley soil (2)	water	$<10^1$		10^2-10^3	10^1-10^3	$<10^2$	2.4–7.4
	soil	10^1-10^2	$<10^1$	10^5	10^3-10^5	10^3-10^4	(4.2)
Low humic gley soil (3)	water	$<10^1$		10^2-10^4	10^1-10^3	10^2-10^3	2.0–9.9
	soil	10^1-10^2	$\leq10^1$	10^5	10^3-10^4	10^3-10^4	(5.1)
Humic gley soil	water	10^1		10^3	10^1	10^3	16.0
	soil	10^4		10^6	10^3	10^4	
Non-calcic brown soil	water	$<10^1$		10^3	10^3	10^2	4.4
	soil	10^3	$<10^1$	10^5	10^3	10^3	
Gray podzolic soil	water	10^2		10^3	10^3	10^4	4.6
	soil	10^3		10^5	10^4	10^4	
Grumusol	water	10^3		10^3	10^2	10^3	8.3
	soil	10^3		10^5	10^3	10^4	
Regosol	water	$<10^0$		10^4	10^1	$<10^0$	0.7
	soil	10^1		10^3	10^2	10^2	

* By acetylene reduction activity measurement
(Source: Matsuguchi et al., 1974)

Table 8.10 Effect of N fertilization and inoculation with Rhizobium leguminosarum bv. Trifolii rice endophytes on production of Giza 175 rice under field condition

Fertilization kg N/ha	Grain yield (t/ha) C	E11	E12	Harvest index (%) C	E11	E12	Fertilizer use efficiency C	E11	E12
0	4.3	6.3	6.2	30.1	35.0	35.4	–	–	–
48	5.8	7.9	7.0	30.0	40.0	38.5	121.7	165.2	145.0
96	5.4	7.1	7.9	31.8	36.7	37.4	56.4	74.1	81.8
144	6.4	7.0	7.0	31.6	30.6	30.0	44.3	48.6	48.4
LSD (0.05)									
Fertilizer N (N)			0.31			2.5			3.4
Inoculation (I)			0.32			1.9			4.4
N × I			0.65			3.9			8.8

(Source: Yanni et al., 1997)

among different microbial groups in the rice rhizosphere. The same authors also mentioned the prospect of endophytic N fixation in the rhizosphere of non-leguminous plants.

8.4 Mineralization of soil organic nitrogen

8.4.1 Mineralization of soil organic nitrogen and soil fertility

As shown in Chapter 7, relatively unhumified organic matter is enriched in paddy soils by repeated yearly submergence. This could serve as a potential source of mineralizable N. Although inorganic forms of N constitute only about 1% of total N before submergence, during the cropping season 1–5% of total N becomes available to rice. Many studies conducted in Japan, including that cited in Table 8.4, showed that even in a field that has sufficient applied fertilizer, N uptake by rice through the mineralization of organic N well exceeds that from fertilizers. This also appears to be true in the tropics. A study conducted by Koyama *et al.* (1973) in Thailand, showed that more than 60% (or 52 kg out of 85 kg) of total N taken up by rice plants by the time of harvest came from mineralization of soil organic N. Thus, the importance of the mineralization process of soil organic N for the growth is obvious.

Organic matter has to be applied to increase the sources of mineralizable N in soil. Shibahara *et al.* (1994 a) reported that organic matter application increased N uptake of rice by 31 kg N ha^{-1} for four years and decreased the percolation loss of N from a paddy field, presumably because of N immobilization, by 37 kg N ha^{-1} over the same period. This is a good way to show that enriching a soil with organic matter not only increases soil N fertility, but also enhances the soil's capacity to suppress environmental pollution.

However, it is important to pay attention to the quality of organic matter. Ito and Iimura (1989) clarified that coarse organic matter > 100 mesh in size amounted to as much as 3.3–7 t ha^{-1} in a paddy field, but made no contribution to N mineralization, which emphasizes the importance of finer fractions of organic matter.

Paddy fields that have been left fallow for a year or two in upland conditions tend to accumulate residues of weeds, so increasing N fertility (Anzai, 1992) to the extent that N fertilization to the succeeding rice crop must take this into account.

The nature and amount of clays are also relevant to N mineralization. Soils with a high clay content, particularly those having clay minerals with a high CEC, tend to show a high amount of N mineralization (Hirokawa and Kitagawa, 1988). The relationship is expressed by the hyperbolic function of $Y = X/(a + bX)$, where Y is the amount of N mineralization and X is a product of total carbon with the sum of smectite and vermiculite in the clay fraction.

The release of mineralized soil organic N generally follows a parabolic curve, while the pattern of N uptake by rice is exponential until the maximum tillering stage, and then becomes linear for the mid to later growth stages (Takahashi et al., 1976). Thus, according to Toriyama (1994, 1996), it is possible to recognize two phases of rice growth, with the division between phase I and phase II at the point of disappearance of NH_4-N liberated by the soil-drying effect (see below). In this case, the main factor influencing the change in annual N uptake during phase I is the soil-drying effect, and that during phase II is the rate of mineralization, as affected by soil temperature. Therefore, in the following section, two stages of soil N mineralization, corresponding respectively to phases I and II of rice growth, are discussed separately.

8.4.2 Nitrogen mineralization as affected by soil-drying prior to submergence

There have been many studies on how soil management factors accelerate soil N mineralization, and the 'soil-drying effect' is probably the most well known and most important. When soil is dried prior to submergence, N mineralization or NH_4-N production is enhanced after submergence. This is called the 'soil-drying effect'. Freezing and treating the soil with a highly concentrated salt solution have an effect similar to soil-drying because dewatering accompanies these treatments. The magnitude of the effect varies between soil types. For example, wet paddy field soils tend to show a greater effect with less of drying, while dry paddy field soils require more drying.

Toriyama (1994) found that when soil was dewatered beyond a certain critical level of moisture content (that is equivalent to a soil moisture tension of pF 4 or 1 MPa), N mineralization was proportional to the amount of dewatering, as expressed by:

$$Y = a(K - X) + Y_0$$

where Y is the amount of N mineralized from the submerged incubation of air-dried soil at 30°C; a, the proportionality constant, and K, the critical moisture content, are intrinsic to each soil; X, is the soil moisture content; and Y_0, is the amount of N mineralized by incubating the soil without drying. Thus, $(Y-Y_0)$ gives the soil-drying effect.

The amount of N mineralized by the soil-drying effect can also be predicted using the amount of N extracted by a pH 7 phosphate buffer solution. It was also reported that the amount of hot water extractable hexoses could be used as a good index. In all these examples, the main source of N measured as the soil-drying effect is considered to be biomass N.

The incubation temperature of a soil sample affects its rate of N mineralization. The difference in the amount of N mineralized between 30°C and 40°C is called the 'soil temperature raising effect'. Generally, this effect is smaller than the soil-drying effect, but these two effects usually occur in parallel. Changes in soil reaction and the mechanical agitation of soil samples are also known to enhance N mineralization.

8.4.3 Nitrogen mineralization as affected by soil temperature after submergence

The mineralization of organic N in the soil is considered to be performed by extra-cellular enzymes, so an enzymatic reaction model may be adopted for the mineralization process. In this model, the reaction follows first order kinetics, and the rate of mineralization is proportional to the substrate concentration. Also, the rate constant of mineralization increases exponentially with a rise in temperature, following the Arrhenius plot: $k = A \cdot \exp(-Ea/RT)$, where k is the reaction rate, A is a constant, Ea is the activation energy, R is the gas constant and T is the absolute temperature. Although many enzymatic reactions are involved, if the plot is linear in a certain temperature range, it can be assumed that one rate-determining reaction dominates that temperature range (Konno, 1980). The activation energy is a parameter related to the effect of temperature on the rate constant of mineralization, and it usually varies in the range of 5000–25000 cal mol^{-1} for enzymatic reactions.

Sugihara et al. (1986) developed a method for interpreting soil N mineralization based on the kinetics of enzyme reaction. They incubated soils under submergence for three to five months at three different temperatures, and fitted the measured NH_4-N data to an equation derived from the first order kinetics:

$$N = N_o \{ 1 - \exp(-kt) \}$$

where N (mg N kg^{-1} dry soil) is the amount mineralized at time t (day), N_o corresponds to readily decomposable N initially present in the soil (mg N kg^{-1} dry soil) and k is the rate constant of mineralization. They further calculated activation energy, Ea, from the plots for three different temperatures. In addition to the simple model given above, they formulated equations consisting of two exponential terms for soils with N fractions that differed in their ease of decomposition.

Using the kinetic approach, the relationships between each of the parameters (N_o, k and Ea) and soil type, and/or the air-drying effect were studied. Yamamoto et al. (1993) indicated that N_o varied for different soil types and increased with the continuous application of organic matter, whereas k is more intrinsic to the soil type and the difference in Ea is relatively small among soil types. Fujii et al. (1989) reported that soils with a high Ea showed a large annual variation in uptake of soil N by rice. Inubushi (1990) and Ueno (1994) noted that analysis of the N mineralization kinetics of dried soils, when incubated under submergence, made it possible to differentiate the rapidly decomposing fraction (N_{oQ}) and the slowly decomposing fraction (N_{oS}). Ando et al. (1989) claimed that for many soils, a two-term model including a negative term could be formulated, enabling them to differentiate N mineralization and immobilization.

Although useful, using experimental procedures of the kinetic approach to predict the N mineralization potential of a soil, is quite involved and time consuming. There is another prediction method for N mineralization that has proved useful as a routine procedure. Yoshino and Dei (1977) proposed a method based on the effective cumulative temperature. They reported that the amount of N released from soils is proportional to the effective cumulative temperature. The mineralizable NH_4-N (Y) may be estimated empirically using the following parabolic equation:

$$Y = k[(T - T_o)D]^n$$

where T is the incubation temperature (or mean soil temperature for the 5 cm depth of a plow layer soil) in °C; T_o is the standard threshold temperature, that is 15°C in Japanese conditions; D is the days of incubation (or any length of time in the field); $(T - T_o)$ D is the effective cumulative temperature; and k and n are constants related to the quantity and pattern of the evolution of NH_4-N, respectively. For example, a low value of n (< 1) corresponds to a pattern of vigorous N release in the early growth stage, while a high value of n (> 1) corresponds to that of a higher rate of N release later in the growth stage. The value of k is related to the N supply potential of a soil. Yoshino and Dei (1977) showed that by using the above equation, based on a laboratory incubation experiment, it was possible to predict the amount of N release for Japanese paddy soils.

The relationship between mineralizable NH_4-N (Y) and the effective cumulative soil temperature for one cropping season of rice is approximated by the following equation:

$$Y = k_{1050}[(T - 15) D/1050]^n$$

where k_{1050} is the amount of soil N mineralized by the effective cumulative soil temperature for the whole cropping season. The figure 1050 is considered to represent the effective cumulative soil temperature of one cropping season of rice in Japan, assuming an effective growing period of ten weeks or seventy days at a soil temperature of 30°C. As this method predicts variations of N mineralization sufficiently well for years of average climate, Miyama (1988) used it as the basis not only for recommending fertilizer management for different types of soil, but also for recommending the dose of top-dressing N.

In both of the above models, only the plow layer of paddy soils is considered. However, what is the contribution of the subsurface and subsoil to N nutrition of rice? Sekiya and Shiga (1975) and Ando et al. (1990) estimated the amount of N taken up by rice from the subsurface and subsoil to be 7–22% of the total uptake. Kaneta (1993) reported a value as high as 65% for the contribution of subsoil in an alternate rice-soybean field, as compared to only 20% in a continuous rice field. This is caused by the deep penetration of rice roots along cracks formed during the upland condition for soybeans. Ando et al. (1990) found in their kinetic studies that N_o, the initial amount of readily mineralizable N, and k, the rate constant, for the subsoil, were about one half that of the plow layer.

8.4.4 Sources of mineralizable N

Repeated drying and wetting to promote the mineralization of the readily decomposable organic N fraction leads to a decrease in acid-hydrolyzable N, especially in amino acid N and amino sugar N. This appears to suggest that the greater part of the readily decomposable organic matter might be protein-like substances. These substances are stabilized or protected from microbial decomposition in the soil through association with organic (humic acid, polyphenols, quinoids, etc.) and inorganic colloids (clay minerals, free Fe and Al hydrated oxides). Various treatments, such as drying and mechanical agitation disrupt the association, resulting in the promotion of N mineralization.

Marumoto et al. (1974) found that when plant residues are added to the soil, their amino acid composition is changed to that intrinsic to the soil

Table 8.11 Distribution of D-amino acids in acid hydrolyzate of paddy soils

Soil	TN % dry soil	D-alanine %	D-glutamic acid, %	Soil texture	Dominant clay minerals
Futsukaichi	0.27	8.8	5.4	SL	hydrated halloysite
Toyama	0.16	11.3	7.6	SL	kaolinite
Kasuya	0.23	10.2	7.5	LiC	kaolinite
Isahaya	0.24	9.8	7.7	LiC	montmorillonite
Miyakonojyo	0.38	14.3	6.7	C	allophane
Hitsujigaoka	0.30	12.7	8.0	C	allophane
Aomori	0.30	10.5	7.5	CL	allophane

Notes: Percentage figures for D-alanine and D-glumatic acid were obtained as follows: (D-alanine/D,L-alanine) × 100; (D-glutamic acid/D,L-Glutamic acid) × 100. The amino acids were contained in acid hydrolyzates of the soil.
(Source: Kawaguchi *et al.*, 1974)

within a relatively short time. Furthermore, the soil amino acid composition becomes similar to the amino acid composition of soil microorganisms, particularly of their cell walls. The amino sugar content of the soil is very high compared with its low content in plants. Kawaguchi *et al.* (1974) also found a specificity of soil amino acid composition that is characterized by the presence of a high proportion of D-amino acids which do not occur in the proteins of animals and plants, but occur in the cell walls of bacteria and actinomycetes (see Table 8.11). D-alanine and D-glutamic acid in the soil are readily decomposed by such treatments as soil-drying and mechanical agitation. Kawaguchi *et al.* (1974) inferred from their data that about 30–60% of soil organic N, and 20–50 % of the readily decomposable N, are derived from microbial cell wall materials. Thus, the contribution of the microbial cells and the structural proteins of animals and plants to the mineralizable N source is quite large.

Today, the major sources of mineralizable N are considered to be microbial biomass and organic matter derived from plant remains, particularly proteins and cell wall material. Inubushi (1984) and Inubushi and Watanabe (1986) differentiated biomass N and non-biomass N by soil incubation, and regarded the sum of both the biomass and non-biomass N, as available N. Biomass N is measured as the NH_4-N liberated by incubating soil after chloroform fumigation treatment, while non-biomass N is that liberated by the incubation of non-treated soil. They monitored the dynamics of both forms of N for a year in paddy soil in the Philippines, and found that 38% of the available N came from biomass N and 62% from non-biomass N. They further estimated that the mean residence time of N in this available N pool was thirty-three days, whereas the periods of N that were evolved as the air-drying effect and from soil humus were seventy-four days and 1460 days, respectively (Inubushi, 1984). Later, Inubushi *et al.* (1991) applied their chloroform fumigation-extraction method, as an improvement of the chloroform fumigation-incubation method, to paddy soils and made possible a more reliable measurement of biomass N. Shibahara *et al.* (1998) further confirmed the applicability of the fumigation-extraction method for estimating the biomass of paddy soils under submerged conditions.

As a result of uptake by rice, sources of available N naturally diminish, but at the beginning of the following cropping season, the amount of readily available N in the soil (as measured with the incubation method) is normally restored to the level of the preceding year (Toriyama and Iimura, 1987). Watanabe and Inubushi (1986) showed an accumulation of biomass N, chlorophyll-like substances, and other components that comprise the soil-drying effect in the surface layer of submerged soils.

8.4.5 Immobilization of applied mineral nitrogen and its remineralization

When mineral N is applied to the soil, some is immobilized by microorganisms and further converted to soil organic N, which is later remineralized. Immobilization and mineralization processes are dependent on many factors, such as temperature, the source of N and carbon, and the C/N ratio. High temperatures accelerate both the immobilization and mineralization processes, while low temperatures (<20°C) retard mineralization considerably, but less so for immobilization (Toriyama and Miyamori, 1988).

NH_4-N appears to be slightly more readily immobilized and less readily remineralized than NO_3-N (Ahmad *et al.*, 1972). Carbon sources that contain more cellulose and lignin immobilize N slowly, and the immobilized N is also remineralized slowly. More carbon is needed to immobilize the same amount of N for those materials with high cellulose and lignin contents (Ahmad *et al.*, 1969). However, such materials accumulate more N in the soil because of the slow remineralization rate (Yoshida *et al.*, 1973). Nitrogen immobilization seems to proceed more readily in the upland condition than in the submerged condition when carbon sources are added (Asami, 1971), but remineralization of the immobilized N also proceeds more readily in the upland condition (Hirose, 1973).

Organic matter with a high C/N ratio immobilizes more N, and the release of the immobilized N is retarded (Ahmad *et al.*, 1969). The newly immobilized N tends to be more readily decomposed than the soil organic matter, but after a rapid decomposition, the remainder is more stable. Dei and Yoshino (1972) used ^{15}N labeled ammonium sulfate as the basal fertilizer in a pot experiment, and found that about 30% of the applied N was recovered at the tillering stage, and about 20% was recovered at the harvesting stage as soil organic N. Furthermore, about 20% of the residual organic N was taken up by rice in the next cropping season. In their experiment, the percentage of applied N that was taken up was 57% in the first crop, 4% in the second, 2% in the third, and 1% in the fourth crop (see also Maeda, 1983).

In paddy soils, an increase in mineral N fertilizer increases the soil organic matter content up to a certain optimal level. However, a dose higher than the optimal decreases the organic matter content. This means that an adequate amount of mineral N fertilizer increases the amount of roots and stubble, and thus the amount of immobilized N. However, an overdose leads to a decrease in the amount of roots and organic N by lowering the C/N ratio of the residues and by the priming effect – the enhancement of the decomposition of native organic matter by an addition of readily available N (Takai and Kitazawa, 1975).

The newly immobilized organic N has a very high percentage of hydrolyzable N, of which amino acid N and unidentified N fractions form a large component. In the remineralization process, these latter two fractions undergo a higher rate of decomposition.

The dynamics of soil and fertilizer N have been studied intensively using the ^{15}N tracer technique and the ^{15}N isotope dilution method, (Yamamuro, 1991; Takahashi and Yamamura, 1994; Ueno, 1994; Shibahara, 1994 b). Nishio et al. (1993), Nishio (1994), and Toriyama et al. (1994) estimated the rates of gross mineralization[2] and immobilization of soil N using the isotope dilution method. Nishio et al. (1993) found that when ^{15}N had been added at the beginning of incubation, some of the once immobilized ^{15}N started to be remineralized after two weeks of incubation. They also found that a plot applied with rice straw showed a higher immobilization in the early stage as compared with a control plot, but after two months of submergence, immobilization decreased and the rate of mineralization became even greater in the plot with added straw than in the control plot.

8.5 Denitrification

8.5.1 Microbial denitrification

Among the causes of the loss of N from paddy soils, denitrification is most important and serious. The magnitude of loss of applied N has been estimated to range from 20–40% in India, and from 30–50% in Japan (De

Datta, 1978). There could be two mechanisms of denitrification, microbial and chemical.

The greater part of denitrification is microbially mediated. As stated earlier, under aerobic conditions ammonia is converted first to nitrite and then to nitrate. Some of the facultative anaerobes utilize NO_3^- as the electron acceptor when the medium becomes slightly anaerobic, thereby reducing it to NO_2^-. The latter is further reduced to N_2 by the denitrifiers, according to the following reaction:

$$NO_3^- + 6H^+ + 5e = (1/2) N_2 + 3H_2O$$

Ito and Iimura (1983, 1984) separately measured denitrification from soil N and from fertilizer N (^{15}N), by measuring the $^{15}N_2$ ratio using GC-MS. They found that about 24% of fertilized N was denitrified in the field, and denitrification from soil N was about 17–20% that of fertilizer N.

Yamamuro (1991) estimated the amount of denitrification for each stage of rice growth using the ^{15}N tracer technique. Denitrification is extensive for the early to mid terms of rice growth, which corresponds to the period when a thin oxidized layer is kept at the soil water interface, and thus a high activity of nitrifiers is maintained.

Nitrous oxide, N_2O, is always emitted as a by-product of denitrification. The relative abundance of this gas varies greatly, depending on so many conditions that at present it is difficult to predict. Generally however, the more the condition is reductive, the less the N_2O evolution, and *vice versa*. (The environmental implications of nitrous oxide are discussed in Chapter 12.)

8.5.2 Chemical denitrification

There are several reactions leading to the loss of N from nitrous acid.
- Nitrous acid is decomposed under acidic condition as follows:

$$3HNO_2 = NO + HNO_3 + H_2O$$

For the reaction to proceed, the medium pH must be below 5.5.
- Nitrous acid undergoes the Van Slyke reaction, with amino acid to liberate N_2 gas:

$$RNH_2 + HNO_2 = ROH + H_2O + N_2$$

- A reaction between nitrous acid and ammonia or urea liberates N_2 gas:

$$NH_3 + HNO_2 = (NH_4NO_2) = N_2 + 2H_2O$$

• Reaction between nitrous acid and reducing organic matter produces volatile gas.

Under ordinary soil conditions, all of the above reactions that lead to chemical denitrification are of no great practical importance.

8.5.3 Factors affecting the denitrification process

There are several factors that may influence denitrification in the soil.

1) Oxygen

Denitrifying bacteria are facultative anaerobes, utilizing oxygen in aerobic conditions and nitrate in anaerobic conditions for respiration. Thus, the low partial pressure of oxygen is a prerequisite for denitrification. However, denitrification can proceed even under a relatively high partial pressure of 2.27%, and loss is enhanced when glucose is added (see Table 8.12). The differentiation of a thin oxidized surface layer of paddy soils provides almost an ideal condition for the loss of N in paddy soils. When organic matter is decomposed by general purpose decomposing bacteria, N is liberated as NH_4^+ (ammonification). If the liberated NH_4-N remains as it is, there is no denitrification. Only when NH_4^+ is nitrified under an aerobic condition in the oxidized layer can denitrification occur. Thus, shallow surface water, adequate percolation and a vigorous growth of algae are conditions that promote the differentiation of the oxidized layer and denitrification.

The proportion of N_2O relative to N_2 is higher in upland soils, where a generally high oxygen partial pressure seems to favor N_2O formation.

2) Organic matter

The presence of organic matter is necessary to provide electrons for the reduction of nitrate. As seen earlier, NO_3^- is more stable in the presence of oxygen than N_2 when the medium has a pH 7:

$$O_2 + 4H^+ + 4e = 2H_2O \qquad\qquad pE^\circ_7 = 13.75$$
$$2NO_3^- + 12H^+ + 10e = N_2 + 6H_2O \qquad\qquad pE^\circ_7 = 12.65$$

Thus, energy is required to reduce NO_3^- to N_2. Organic matter supplies this energy as it is being decomposed. The second role of organic matter in the denitrification process is to consume oxygen in the medium, thus facilitating the reduction of NO_3^-. The effect of glucose in Table 8.12 (Allison and Doetsch, 1950) is explained by these two actions.

Table 8.12 Effects of oxygen and glucose on the loss of nitrogen

Incubation period, days	No glucose				Glucose added			
	NO_3-N		Total N		NO_3-N		Total N	
	mg found	% lost	mg found	mg lost	mg found	% lost	mg found	mg lost
			0.46% Oxygen					
0	78.3		154.7		78.3		154.7	
5	72.5	7.3	147.9	6.8	23.2	70.4	150.8	3.9
10	63.5	18.8	143.4	11.3	1.4	98.2	114.8	39.9
			2.27% Oxygen					
0	78.3		154.7		78.3		154.7	
5	78.0	1.0	149.0	5.7	55.5	29.1	150.5	4.2
10	76.5	2.2	152.4	2.3	58.5	25.2	143.6	11.1

Notes: Labeled KNO_3 was added to 20g of soil (TN: 0.06–0.07%, TC: 0.84%, pH 5.6) in an Erlenmeyer flask. Glucose was added to make its content 0.5%. Results are expressed for 100 g soil.
(Source: Allison and Doetsch, 1950)

Araragi and Tangcham (1974) studied the effect of organic matter on denitrification in Thailand. Table 8.13 shows that in the plot with added rice straw (containing 30kg N), there was a loss of 18.6 kg N during the two month period from transplanting. This was nearly ten times more than the N loss from the control plot.

Toriyama (personal communication) reviewed denitrification studies that had been conducted in Japan with the use of ^{15}N and found no case of enhancement of denitrification of basal N when rice straw was added, although there were a few studies reporting increased denitrification of top-dressed N in the plot with rice straw. In view of this, the results presented in Table 8.13 deserve special attention. The enhancement of denitrification of basal N by rice straw application may be a phenomenon specific to tropical soils with a low organic matter reserve in their original state. Further studies are needed to confirm the data in Table 8.13.

3) pH
Denitrification is influenced by soil pH, and generally, it increases with increasing pH. This is because the growth of bacteria involved in de-nitrification is optimal at, or near, neutrality. It may be of some importance that the decomposition of organic matter, and accordingly the consumption of oxygen, are promoted in the neutral pH range.

4) Temperature
The optimum temperature for denitrification is reported as 25°C or a little

Table 8.13 Effect of organic matter on denitrification

No.	Treatment	Days after Transplanting (TP)						
		7	7–13	13–20	2–34	34–49	49–64	Sum
1	Control	30	26	28	51	12	29	1.9
2	P	56	166	37	66	37	25	3.5
3	$(NH_4)_2SO_4$ +P	94	84	17	27	20	17	2.2
4	NH_4Cl +P	94	140	37	42	28	15	3.0
5	Amorphous P	91	106	63	90	37	38	4.1
6	2) + straw	360	256	168	365	524	28	18.6
7	3) + straw	411	779	205	379	631	261	24.6
8	4) + straw	246	646	155	175	257	193	15.9
9	5) + straw	360	569	157	328	243	178	18.3

Notes: (1) Figures are N in g ha^{-1} d^{-1}; Sum in N kg ha^{-1} (2) Soil (CL, TC 0.43%, avail.P$_2$O$_5$ 21.1mg kg^{-1}, avail.K$_2$O 316 mg kg^{-1}, pH 6.6–7.1), rice straw (applied 2 weeks before TP) 6,000 kg, N 32 kg, P$_2$O$_5$, 40 kg, K$_2$O 40 kg.
(Source: Araragi and Tangcham, 1974)

higher. Temperature affects the relative amounts of N$_2$O and N$_2$, the latter becoming more dominant at higher temperatures.

5) *Eh* (redox potential)
Pearsall and Mortimer (1939) reported that NO$_3^-$ accumulated when the *Eh* was above 350 mV, while it disappeared as the *Eh* went below 320 mV. Patrick (1960) reported that NO$_3^-$ becomes unstable when the *Eh* goes below 338 mV. He further stated that the *Eh* is more relevant to denitrification than partial pressure of oxygen. However, Keafauver and Allison (1957) reported that the *Eh* was not a key factor for nitrite reduction.

8.5.4 Denitrifying microorganisms
All the denitrifying microorganisms known so far are bacteria and most of them are heterotrophs. There are however, some autotrophs, such as *Micrococcus denitrificans* and *Thiobacillus denitrificans*. The former utilizes nitrate anaerobically and draws energy from the oxidation of hydrogen, while the latter lives on sulfur oxidation, reducing nitrate simultaneously.

8.5.5 Counter-measures to denitrification
It is known from the mechanism of denitrification that if ammoniacal N is not nitrified at the soil-water interface, there will be no denitrification. Therefore, one effective counter-measure would be to place fertilizer N deep into the soil. Chemicals retarding nitrification (nitrification inhibitors, such as N-serve®), granulated fertilizers with small surface area for microbial

transformation, (such as urea super-granule) and avoiding the simultaneous application of mineral N fertilizer and organic matter, could also have some effect. However, deep placement seems to be the easiest and the best counter-measure. Fertilizer applicators for this purpose are now available.

Recently, use of plastic-coated urea preparation is becoming increasingly popular for rice cultivation in Japan. As is elaborated in Chapter 12, this may be considered a counter-measure to denitrification, as it dramatically raises the use efficiency of fertilized N.

8.6 Ammonia volatilization

The loss of gaseous ammonia from paddy soils has not received much attention. Of course, in a soil with an alkaline reaction (say, pH 8.5) ammonia volatilization can be quite significant. Iwata and Okuda (1937) reported in one experiment that 17.4% of the applied N was lost from soil with pH of 8.5. However, in most paddy soils with acidic reaction, this source of N loss has been considered insignificant.

Wetselaar et al. (1977) reported that the loss of N, applied as ammonium sulfate to the surface of paddy soil near Chai Nat, Thailand, was as high as 12%. The soil pH in this case was between 7.0 and 7.5. They also noted that some of the volatilized ammonia was directly assimilated by rice plants. Mikkelsen et al. (1978) found that the pH of shallow flood water was greatly affected by the total respiratory activity of all the heterotrophic organisms and the gross photosynthesis of the species present. The pH increased by midday to values as high as pH 9.5–10 and fell by as much as 2–3 pH units during the night. Thus, NH_4^+ fertilizer broadcast into a high pH water could be highly susceptible to direct NH_3 volatilization losses. Mikkelsen et al. also showed that N losses on fertile, neutral pH Maahas clay in the Philippines were as high as 20% of the amount applied.

Today, loss of fertilizer N due to ammonia volatilization is considered significant in tropical regions, where high temperatures and strong sun-light greatly reduce dissolved CO_2 in surface water through the enhanced photosynthetic activity of aquatic plants in the middle of the day. Similar diurnal fluctuations of pH in shallow surface water and river water have been noted in Japan, but N loss due to ammonia volatilization from paddy fields is still not considered to be very significant.

8.7 Leaching loss of nitrogenous compounds

Today, the leaching loss of N is not only an economic concern, but also a problem of environmental importance. Earlier studies conducted in Japan

Table 8.14 Leaching loss measured in lysimeter experiment (planted with rice)

Soil (year of expt.)	Element	Fertilizer (F) kg ha⁻¹	Leaching loss with F kg ha⁻¹	Leaching loss without F kg ha⁻¹	LLR (leaching loss rate) %	Soil type (texture)
Nishigahara (1928)	N	175	55	23	18.3	Andosol (L)
	P	69	1	1	0	
	K	133	26	24	1.2	
Konosu (?)	N	132	10	8	1.5	Alluvial (?)
Aichi (1931)	N	185	81	34	25.4	Alluvial (CL)
	P	81	0.1	0.1	0	
	K	154	26	23	2.2	
Shimane (1962)	N	100	32	28	3.7	Semi-wet paddy (L)
Kuroda (1968)	N	300	28	2	8.7	Alluvial (CL)
Tsu (1968)	N	300	52	1.4	12.7	Alluvial (CL)
Matsusaka (1968)	N	300	60	18	14	Alluvial (CL)
Shiga (1972)	N	100	30	21	9.0 (6.9)*	Alluvial (L)
	P	88	2.4	1.3	1.3	

* LLR during the cropping period in parentheses
(Sources: Okuda, 1949; Kanei and Matsuda, 1931; Yamane and Matsuura, 1963; Tokai-Kinki Agr. Expt. Sta., 1969; Nakada *et al.*, 1976)

were mainly from the viewpoint of N economy, while recent studies are mainly related to the eutrophication of water bodies.

As it is difficult to collect percolation water in the field, experiments using lysimeters have been adopted to study the leaching loss of N. Usually, the estimates obtained by lysimeter studies are larger than those expected for actual field conditions, because of the seepage along the lysimeter walls and the absence of surface run-off. Table 8.14 summarizes the results of lysimeter studies conducted to estimate the leaching loss of fertilizer N. The leaching loss rate, LLR, is calculated from the following formula:

LLR = (leaching loss from fertilized plot – leching loss from the non-fertilized plot) / applied N

Table 8.15 shows the distribution of applied N tagged with ^{15}N among rice, soil, and percolation water (outflow). In this particular study, the losses due to denitrification (unrecovered) seem to be unimportant, and this is probably due to the low organic matter reserve of the soil.

Generally, leaching loss is more significant for the fertilizer N applied during the earlier stages of rice growth. The very insignificant leaching loss from the top-dressed N, as in Table 8.15, is not an exception but rather a general rule. This is also shown in Figure 8.4, which indicates low concentrations of N in percolation water for the top-dressed N compared

Table 8.15 Distribution of N (labeled with ^{15}N) applied to rice at Aichi Agricultural Experiment Station (1974)

Whereabouts	Basal application		Top dressing		Sum	
	kg 10a^{-1}	%	kg 10a^{-1}	%	kg 10a^{-1}	%
Rice	1.254	26.1	2.148	79.6	3.402	45.3
Soil	2.784	58.0	0.365	13.5	3.149	42.0
Outflow	0.502	10.5	0.004	0.2	0.506	6.8
Unrecovered	0.260	5.4	0.183	6.7	0.443	5.9
Sum	4.8	100.0	2.7	100.0	7.5	100.0

(Source: Asano and Kanda, 1976)

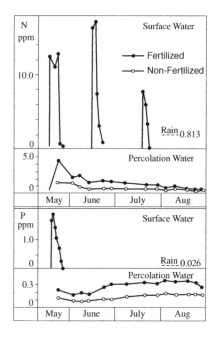

Figure 8.4
Nitrogen and phosphorus concentrations in surface water and percolation water in lysimeter experiment with a rate of percolation of 10mm per day (Source: Nakada, 1972)

with the basal dressing. This may be partly explained by the higher percolation rate at the earlier stage of rice growth due to lower evapotranspiration. However, the main reason could be a rapid uptake of top-dressed N through the well-developed root system of the rice plants at the time of top-dressing.

Notes

(1) List of free-living (or non-symbiotic) N fixing microorganisms
(Source: Matsuguchi, 1978, partly modified)

I. Bacteria (Buchanan and Gibbons, 1974)*

---

(1) Photosynthetic bacteria
Rhodospirillaceae .... *Rhodospirillum, Rhodopseudomonas,*
Rhodomicrobium
Chromatiaceae .... *Chromatium, Ectothiorhodospira, Triospirillum*
Chlorobiaceae .... *Chlorobium, Chloropseudomonas*
(2) Gram-negative aerobic bacteria
Azotobacteriaceae .... *Azotobacter, Azotomonas, Beijerinckia, Derxia*
Pseudomonadaceae ..... *Pseudomonas (P. azotogensis)*
(3) Gram-negative facultative anaerobic bacteria
Enterobacteriaceae ...... *Klebsiella (K. pneumoniae), Enterobacter (E.*
cloecae), Escherichia (E. intermedia),
Flavobacterium sp.
(4) Gram-negative anaerobic bacteria
.... *Desulfovibrio (D. vulgaris, D. desulfuricans)*
(5) Methane forming bacteria
Methanobacteriaceae .. *Methanobacterium, Methanobacillus*
(6) Spore-forming bacteria
Bacillaceae .... *Bacillus (B. polymycxa, B. macerans, B.*
circulans), Clostridium (C. pasteurianum, C.
butyricum), Desulfotomaculum sp.
(7) Bacteria analogous to Actinomycetes
Mycobacteriaceae .... *Mycobacterium (M. flavum)*

II. Blue-green algae (Prescott, 1970)

---

(1) Heterocyst-forming blue-green algae
Nostocaceae .... *Anabaena, Anabaenopsis, Aphnizomenon,*
Aulosira, Chlorogloeopsis, Cylindrospermum,
Nostoc
Stigonemataceae .... *Hapalosiphon, Mustigocladus, Stigonema*
Scytonemataceae .... *Microchaete, Scytonema, Tolypothrix*
Rivulariaceae .... *Calothrix*
(2) Non-heterocyst-forming blue-green algae
Chloococcaceae .... *Anacystis, Aphanothece, Gloecapsa, Gloeothece,*
Microcystis
Entophysalidaceae .... *Chlorogloea*
Oscillatoriaceae .... *Lyngbya, Oscillatoria, Phormidium,*
Trichodesmium
Scytonemataceae .... *Plectonema***

---

* According to Holt *et al.* (1994), *Azomonas (Azotobacteriaceae), Azospirillum* and
Xanthobacter (Rhizobiaceae), and *Herbaspirillum* (others/genera of families *incertae*
sedis), are added to the list. *Azomonas* may be a renaming of *Azotomonas* in the table.
** According to Fogg *et al.* (1973) this belongs to *Oscillatoriaceae*.

(2) The term *gross* mineralization was distinctively used from *net* mineralization, which is measured as the difference between *gross* mineralization and immobilization. Ordinary measurement without the use of ^{15}N can only give *net* mineralization.

References

Ahmad, Z., Kai, H. and Harada, T. 1969. Factors affecting immobilization and release of nitrogen in soil and chemical characteristics of the nitrogen newly immobilized II. Effect of carbon sources on immobilization and release of nitrogen in soil. *Soil Sci. Plant Nutr.*, 15: 252–258.

Ahmad, Z., Kai, H. and Harada, T. 1972. Effect of forms of nitrogen on immobilization and release of nitrogen in soil. *J. Fac. Agric. Kyushu Univ.*, 17: 49–65.

Allison, F.E. and Doetsch, J.H. 1950. Nitrogen gas production by the reaction of nitrites with amino acids in slightly acidic media. *Soil Sci. Soc. Amer. Proc.*, 15: 163–166.

Ando, H., Fujii, H. and Nakanishi, M. 1990. The mineralization pattern of organic N in subsoil in Shonai. *Jpn. J. Soil Sci. Plant Nutr.*, 61: 466–471. (In Japanese).

Ando, H., Fujii, H., Sato, T., Aragaki, K., Nakanishi, M. and Sato, Y. 1989. Models for predicting nitrogen mineralization of paddy soils. *Jpn. J. Soil Sci. Plant Nutr.*, 60: 1–7. (In Japanese).

Anzai, T. 1992. Strategies of nitrogen fertilization for rice in different types of paddy soils restored from fallowing. *Res. Rep. Chiba Agric. Expt. Sta.*, 33: 9–25. (In Japanese).

Araragi, M. and Tangcham, B. 1974. *Report of the Joint Research Work between Thailand and Japan*. Trop. Agric. Res. Center, Tsukuba.

Arihara, J. 2000. *Cropping Systems and Their Mechanisms of Nutrient Uptake*. FFTC Extension Bulletin, 491: 1–18.

Asami, T. 1971. Immobilization and mineralization of nitrogen compounds in paddy soils (Part 4). Effect of plant residues on the immobilization and mineralization of nitrogen compounds in the paddy soil incubated under submerged or upland condition. *J. Sci. Soil & Manure, Japan*, 42: 97–102. (In Japanese).

Asano, M. and Kanda, S. 1976. In Yatazawa, M. (Ed.) *Pollution of Terrestrial Water by Fertilizers and Indicator Plants*. pp. 17–24.

Asano, J. and Yatazawa, M. 1976. In Yatazawa, M. (Ed.) *Pollution of Terrestrial Water by Fertilizers and Indicator Plants*. pp. 17–24.

Becking, J.H. 1976. Contribution of plant-algae association. *Proc. 1ˢᵗ Int. Symp. on Nitrogen Fixation*, 2: 556–580.

Buchanan, R.E. and Gibbons, N.E. (Eds.) 1974. *Bergey's Manual of Determinative Bacteriology, 8ᵗʰ Ed.* Williams and Wilkins Co., Baltimore. (Cited from Kawaguchi, K. (Ed.) 1978. *Paddy Soil Science.* Kodansha, Tokyo: In Japanese).

De, P.K. 1939. The role of blue-green algae in nitrogen fixation in rice fields. *Proc. Roy. Soc. (London)*, B127: 121–139. (Cited from Kawaguchi, K. (Ed.) 1978. *Paddy Soil Science.* Kodansha, Tokyo. In Japanese).

De Datta, S.K. 1978. Fertilizer management for efficient use in wetland rice soils. In IRRI (Ed.) *Soils and Rice*, pp. 671–701. IRRI, Los Baños, Philippines.

Dei, Y. 1975. Accumulation and decomposition of organic matter in paddy soils. *J. Sci. Soil & Manure, Japan*, 46: 251–254. (In Japanese).

Dei, Y. and Yoshino, T. 1972. Utilization of ¹⁵N in paddy soil research. *Nat. Inst. Agric. Sci. Disc. Paper* No. 2, pp. 1–16. (In Japanese).

Denmead, O.T., Freney, J.R. and Simpson, J.R. 1976. A closed ammonia cycle within a plant canopy. *Soil Biol. Biochem.*, 8: 161–164. (Cited from Watanabe, 1978).

Dommergues, Y., Balandreau, J., Rinaudo, G., and Weinhard, P. 1973. Non-symbiotic nitrogen fixation in the rhizospheres of rice, maize and different tropical grasses. *Soil Biol. Biochem.*, 5: 83–89.

Eady, R.R. and Postgate, J.R. 1974. Nitrogenase. *Nature*, 249: 805–810. (Cited from Kawaguchi, K. (Ed.) 1978. *Paddy Soil Science.* Kodansha, Tokyo. In Japanese).

Eady, R.R., Smith, B.E., Cook, K.A. and Postgate, J.R. 1972. Nitrogenase of *Klebsiella pneumonicae*, purification and properties of the component proteins. *Biochem. J.*, 128: 655–675. (Cited from Kawaguchi, K. (Ed.) 1978. *Paddy Soil Science.* Kodansha, Tokyo: In Japanese).

Fogg, G.E. Stewart, W.E.P., Fay, P. and Walsby, A.E. 1973. *The Blue-Green Algae*. Academic Press, N.Y. (Cited from Kawaguchi, K. (Ed.) 1978. *Paddy Soil Science.* Kodansha, Tokyo: In Japanese).

Fujii, H., Ando, H., Sato, T., Aragaki, K., Nakanishi, M. and Sato, Y. 1989. The nitrogen mineralization patterns of paddy soils in Shonai District. *Jpn. J. Soil Sci. Plant Nutr.*, 60: 8–14. (In Japanese).

Hanawall, R.B. 1969. Environmental factors influencing the sorption of atmospheric ammonia by soils. *Soil Sci. Soc. Amer. Proc.*, 33: 231–232. (Cited from Watanabe, 1978).

Hasegawa, K. 1992. Studies on the dynamics of nitrogen in paddy soils and its environmental impact. *Spec. Res. Rep. Shiga Agric. Expt. Sta.*, 17: 1–164. (In Japanese).

Hauck, R.D. 1971. Quantitative esstimates of nitrogen-cycle processes, concepts and review. *Nitrogen-15 in Soil-Plant Studies.* IAEA, Vienna.

Hirokawa, T. and Kitagawa, Y. 1988. Effect of clay composition on mineralization of organic nitrogen in paddy soils. *Jpn. J. Soil Sci. Plant Nutr.*, 59: 41–46. (In Japanese).

Hirose, H. 1973. Mineralization of organic nitrogen of various plant residues in the soil under upland condition. *J. Sci. Soil & Manure, Japan*, 44: 157–163. (In Japanese).

Holt, J.G., Krieg, N.R., Sneath, P.H.A., Staly, J.T. and Williams, S.T. (Ed.) 1994. *Bergey's Manual of Determinative Bacteriology, 9th Ed.* Williams and Wilkins Co., Baltimore.

Hutchinson, G.L., Millington, R.J. and Peters, D.B. 1972. Atmospheric ammonia absorption by plant leaves. *Sci.*, 175: 771–772. (Cited from Watanabe, 1978).

International Rice Research Institute (IRRI) 1992. IRRI Program Report, *Integrated Nutrient Management*, pp. 76–82.

Inubushi, K. 1984. Readily decomposable organic nitrogen in paddy soils. *Soil and Microorganisms*, 26: 31–39. (In Japanese).

Inubushi, K. 1990. Mineralization of soil nitrogen–Trend and prospect of research. In Jpn. Soc. Soil Sci. Plant Nutr. (Ed.) *Nitrogen Mineralization in Paddy Soils and Fertilizer Application*, pp. 9–33. Hakuyusha, Tokyo. (In Japanese).

Inubushi, K. and Watanabe, I. 1986. Dynamics of available nitrogen in paddy soils. II. Mineralized N of chloroform-fumigated soil as a nutrient source for rice. *Soil Sci. Plant Nutr.*, 32: 561–577.

Inubushi, K., Brookes, P.C. and Jenkinson, D.S. 1991. Soil microbial biomass C, N and ninhydrin-N in aerobic soils measured by the fumigation-extraction method. *Soil Biol. Biochem.*, 23: 737–741.

Ishizawa, S., Suzuki, T. and Araragi, M. 1970. Trend of free-living nitrogen fixers in paddy field. *Proc. 2nd Symp on Nitrogen Fixation and Nitrogen Cycle* (Record of Activities of JIBP).

Ito, J. 1977. Behaviour and fixation of nitrogen in paddy field. *Niigata Agronomy*, 13: 51–61. (In Japanese).

Ito, J. and Iimura, K. 1983. Direct measurement of N_2 gas denitrified from soil and fertilizer nitrogen in paddy field using GC-MS. *Jpn. J. Soil Sci. Plant Nutr.*, 54: 235–240. (In Japanese).

Ito, J. and Iimura, K. 1984. Direct measurement of N_2 denitrified from soil and fertilizer in paddy fields using ^{15}N as a tracer. *Japan Agric. Res. Quarterly (JARQ)*, 18: 27–30.

Ito, J. and Iimura, K. 1989. Amount of crude organic matter in fine gley paddy soil in Hokuriku District and its effect on nitrogen mineralization. *Jpn. J. Soil Sci. Plant Nutr.*, 60: 56–59. (In Japanese).

Iwata, T. and Okuda, A. 1937. Volatilization of ammonia from the soil. *J. Soc. Soil and Manure, Japan*, 11: 185–187 (Miscellaneous report). (In Japanese).

Kanai, M. and Matuda, M. 1931. The losses of fertilizing constituents from soils (Part 1). *J. Sci. Soil & Manure, Japan*, 5: 19–40. (In Japanese).

Kaneta, Y. 1993. Studies on elucidation of the effect of land use alternation from upland field to paddy and on stabilized high-yielding techniques for rice in heavy clayey lowland swampy soils in Hachirogata polder land. *Res. Rep. Akita Agric. Expt. Sta.*, 33: 1–45. (In Japanese).

Kawaguchi, S., Kai, H. and Harada, T. 1974. Amino acids in the soil (Part 4). Distribution of D-amino acids in the soil and the significance of bacterial cell wall materials as the source of soil organic matter. *Abstrats of the 1974 Branch Meetings, Soc. Sci. Soil & Manure,Japan*, 21 (Part I): 87. (In Japanese).

Kawasaki, I. 1953. *Natural Supply Potentials of the Three Major Elements in the Main Arable Lands of Japan.* Nippon Nogyou Kenkyusho, Tokyo. (In Japanese).

Keafauver, M. and Allison, F.E. 1957. Nitrite reduction by *Bacterium denitrificans* in relation to oxidation-reduction potential and oxygen tension. *J. Bact.*, 73: 8–14.

Kobayashi, J. 1971. *Health Check of Water.* Iwanami-Shoten, Tokyo. (In Japanese).

Konno, T. 1980. Soil biological activity and temperature. *Soil Phys. Cond. and Plant Growth*, 41: 7–16. (In Japanese).

Koyama, T.,Chammek, C. and Niamsrichand, N. 1973. *Nitrogen application technology for tropical rice as determined by field experiments using ^{15}N tracer technique.* Tech. Bull. (Trop. Agric. Res. Center, Tokyo), No. 3.

Ladha, J.K., Tirol-Padre, A., Reddy, K. and Ventura, W. 1993. Prospects and problems of biological nitrogen fixation in rice production: A critical assessment. In Palacios, S., Mora, J. and Newton, W.E. (Eds.) *New Horizons in Nitrogen Fixation*, pp.677–682. Fluwer Academic Publishers, Dordrecht, The Netherlands. (Cited from Arihara, 2000).

MacRae, I.C. 1974. Effects of applied nitrogen upon acetylene reduction in rhizosphere. Soil Biol. Biochem., 7: 337–338.

Maeda,K. 1983. Quantitative estimation of behavior of nitrogen applied to paddy field. *Bull. Nat. Agric.Res. Center.*, 1: 121–192. (In Japanese).

Marumoto, T., Furukawa, K., Yoshida, T., Kai, H., Yamada, Y. and Harada, T. 1974. Contribution of microbial cells and cell walls to an accumulation of soil organic matter becoming decomposable in soil through the effect of drying a soil (Part 1). Alteration of the contents of individual amino acids and amino sugars contained in the organic nitrogen in soil through the decomposition of rye-grass applied. *J. Sci. Soil & Manure, Japan*, 45: 23–28. (In Japanese).

Matsuguchi, T. 1978. 6.2 Nitrogen fixation in paddy soils. In Kawaguchi, K. (Ed.) 1978. *Paddy Soil Science.* Kodansha, Tokyo. (In Japanese).

Matsuguchi, T., Tangcham, B and Somchai, S. 1974. Free-living nitrogen fixers and acetylene reduction in tropical rice fields. *Japan Agric. Res. Quarterly.(JARQ)*, 8: 253–256.

Matsuguchi, T., Tangcham, B and Somchai, S. 1977. *Ecology of non-symbiotic nitrogen fixers and acetylene reduction in paddy fields of Thailand.* Tech. Bull. (Trop. Agric. Res. Center, Tokyo) No. 6.

Mikkelsen, D.S., De Datta, S.K. and Obcemea, W.N. 1978. Ammonia volatilization losses from flooded rice soils. *Soil Sci. Soc. Amer. J.*, 42: 725–730.

Miyama, M. 1988. Studies on the new method to recommend nitrogen fertilization based on the optimal quantity of nitrogen uptake by rice. *Spec. Res. Rep. Chiba Agric. Expt. Sta.*, 15: 1–92. (In Japanese).

Nakada, H. 1972. A survey on outflows of nutrients from agricultural lands. In Civil Engineer. Soc. (Ed.) *Studies on Future Water Quality of Lake Biwa.* (In Japanese).

Nakada, H., Kawamura, S., Sawa, S. and Katsuki, Y. 1975. Behavior of nutrient salts in crop lands (Part 1). Loss of nutrient salts due to overflowing and leaching. *Abstracts of the 1974 and 1975 Branch Meetings, Soc. Soil and Manure, Japan*, 22 (Part I): 78. (In Japanese).

Nakamura, M. 1970. Biochemistry of nitrogen fixation. *Chem. & Biol.*, 8: 720–731. (In Japanese).

Nishio, T. 1994. Estimating nitrogen transformation rates in surface aerobic soil of a paddy field. *Soil Biol. Biochem.*, 26: 1273–1280.

Nishio, T., Sekiya, H., and Kogano, K. 1993. Estimate to nitrogen cycling in ^{15}N-amended soil during long-term submergence. *Soil Biol. Biochem.*, 25: 785–788.

Okuda, A. 1949. *Introduction to Fertilizer Science.* Yokendo, Tokyo. (In Japanese).

Ono, S. and Koga, H. 1984. Natural nitrogen accmulation in paddy soil in relation to nitrogen fixation by blue-green algae. *Jpn. J. Soil Sci. Plant Nutr.*, 55: 465–470. (In Japanese).

Patrick, W.H. 1960. Nitrate reduction in a submerged soil as affected by redox potential. *Trans. 7th Int. Congr. Soil Sci.*, 2: 494–500.

Pearsall, W.H. and Mortimer, C.H. 1939. Oxidation-reduction potentials in waterlogged soils, natural waters and muds. *J. Ecol.*, 27: 483–501. (Cited from Kawaguchi, K. (Ed.) 1978. *Paddy Soil Science*. Kodansha, Tokyo. In Japanese).

Prescott, G.W. 1970. *The Freshwater Algae*. W.M. Brown Pub. Co., Iowa. (Cited from Kawaguchi, K. (Ed.) 1978. *Paddy Soil Science*. Kodansha, Tokyo. In Japanese).

Quispel, A. 1974. General Introduction. In *The Biology of Nitrogen Fixation*, North-Holland Res. Monographs, Vol. 33, pp. 1–8. (Cited from Kawaguchi, K. (Ed.) 1978. *Paddy Soil Science*. Kodansha, Tokyo. In Japanese).

Sekiya, S. and Shiga, H. 1975. Nitrogen uptake patterns of rice plant grown in Hokkaido. *J. Sci. Soil & Manure, Japan*, 46: 280–285. (In Japanese).

Shibahara, F., Kawamura, M. and Kobayashi, M. 1994 a. Influences of straw incorporation in rotational paddy fields on the generated load of eutrophic materials and soil microorganisms–The effect of rice straw incorporation in fallow paddy fields on the reduction in the discharge of nitrate nitrogen. *Res. Rep. Shiga Agric. Expt. Sta.*, 35: 1–18. (In Japanese).

Shibahara, F., Takehisa, K. and Hasegawa, K. 1994 b. Analysis of nitrogen dynamics in paddy fields applied with organic matter with the ^{15}N tracer technique. *Res. Rep. Shiga Agric. Expt. Sta.*, 35: 105. (In Japanese).

Shibahara, F., Yamamuro, S. and Inubushi, K. 1998. Dynamics of microbial biomass nitrogen as influenced by organic matter application in paddy fields. *Soil Sci. Plant Nutr.*, 44: 167–178.

Sugihara, S., Konno, T. and Ishii, K. 1986. Method of kinetic analysis of organic nitrogen mineralization in soil. *Bull. Nat. Inst. Agro-Environ. Sci.*, 1: 127–166. (In Japanese).

Suzuki, A. 1997. *Fertilization of Rice in Japan*. Japan FAO Association, Tokyo.

Takahashi, Shigeo, Wada, G. and Shoji, S. 1976. Dynamics of soil nitrogen in paddy field and nitrogen uptake by rice. Part 6. Effect of temperature on nitrogen uptake of rice and fate of ammoniacal nitrogen in soil. *Jpn. J. Crop Sci.*, 45: 213–219. (In Japanese).

Takahashi, Shigeru and Yamamuro, S. 1994. Effect of air-drying before flooding on nitrogen mineralization. *Jpn. J. Soil Sci. Plant Nutr.*, 65: 165–170. (In Japanese)

Takai, Y. and Kitazawa, Y. 1975. Soil as an ecosystem. *Kagaku (Sci.)*, 45: 578–585. (In Japanese).

Takamura, Y. and Tabuchi, T. 1977. Outflows of fertilizers from paddy fields and eutrophication of terrestrial water bodies. *Almanac of Water Quality of Rivers, 1977 Ed.*, pp. 861–871, Japan Assoc. Rivers, Tokyo. (In Japanese) (Cited from Kawaguchi, K. (Ed.) 1978. *Paddy Soil Science*. Kodansha, Tokyo. In Japanese).

Tokai-Kinki Agricultural Experiment Station 1969. Research results on soils and fertilizers. (In Japanese) (Cited from Kawaguchi, K. (Ed.) 1978. *Paddy Soil Science*. Kodansha, Tokyo. In Japanese).

Toriyama, K. 1994. Studies on estimation of nitrogen mineralization pattern of lowland rice field and nitrogen fertilizing model for rice plant. *Bull. Hokuriku Agric. Expt. Sta.*, 36: 147–198. (In Japanese).

Toriyama, K. 1996. Progress and prospect of the research on paddy soil management under various rice growing systems: 1. Progress in nutrient behavior and management research on paddy soil (1) Nitrogen. *Jpn. J. Soil Sci. Plant Nutr.*, 67: 198–205. (In Japanese).

Toriyama, K. 2002. Estimation of fertilizer nitrogen requirement for average rice yield in Japanese paddy field. *Soil Sci. Plant Nutr.*, 48: 293–300.

Toriyama, K. and Iimura, K. 1987. Seasonal changes of mineralizable nitrogen at clayey gley paddy field in Hokuriku District of Japan. *Jpn. J. Soil Sci. Plant Nutr.*, 58: 671–676. (In Japanese).

Toriyama, K. and Miyamori, Y. 1988. Estimation of gross nitrogen mineralization and immobiliation rate in submerged paddy soil by ^{15}N isotope dilution technique. *Jap. J. Soil Sci. Plant Nutr.*, 59: 56–60. (In Japanese).

Ueno, M. 1994. Desirable growth and nitrogen uptake patterns for stable production of high quality rice (cultivar: *Sasanishiki*) and development of nitrogen fertilization method taking soil nitrogen into consideration in Yamagata Prefecture. *Spec. Res. Rep. Yamagata Agric. Expt. Sta.*, 22: 1–86. (In Japanese).

Van Breemen, N., Burrough, P.A., Verthorst, E.J., van Dobben, H.F., de Wit, T., Ridder, T.B., and Reijnders, H.F.R. 1982. Soil acidification from atmospheric ammonium sulphate in forest canopy throughfall. *Nature* (London), 299: 548–550.

Watanabe, I. 1978. Biological nitrogen fixation in rice soils. In IRRI (Ed.) *Soils and Rice*, pp. 465–478. IRRI, Los Baños, Philippines.

Watanabe, I. and Inubushi, K. 1986. Dynamics of available nitrogen in paddy soil. I. Changes in available N during rice cultivation and origin of N. *Soil Biol. Biochem.*, 27: 37–50.

Watanabe, I., Lee, K.K., Akimagno, B.V., Sato, M., Del Rosario D.C. and De Guzman, M.R. 1977. *Biological Nitrogen Fixation in Paddy Field by in situ Acetylene-Reduction Assays*. IRRI, Res. Pap. Ser. 3, Los Baños, Philippines.

Wetselaar, R., Shaw, T., Firth, P., Oupathum, J. and Thitipoca, H. 1977. Ammonia volatilization losses from variously placed ammonium sulphate under lowland rice field conditions in central Thailand. *Proc. Int. Seminar Soil Environ. Fertility Management in Intensive Agric. (SEFMIA)*, pp. 282–288.

Yamamoto, T., Tanaka, K. and Kakushige, K. 1993. Pattern of soil nitrogen mineralization and diagnosis of fertilization for paddy soils in the warm climatic region. Part 2. Characteristics of soil nitrogen mineralization in paddy soils and nitrogen uptake pattern during the rice growing period. *Jpn. J. Crop Sci.*, 63: 363–371. (In Japanese).

Yamamuro, S. 1987. Nitrogen cycle in paddy fields (Part 9). Method for measuring the amount of biological nitrogen fixation in paddy fields using a N_2-^{15}N dilution technique. *Jpn. J. Soil Sci. Plant Nutr.*, 58: 444–451. (In Japanese).

Yamamuro, S. 1991. Theoretical approach towards ^{15}N tracer method for measuring mineralization, immobilization, denitrification and uptake in paddy fields, and prediction of nitrogen taken up by rice. *Bull. Kyushu Nat. Agric. Expt. Sta.*, 27: 1–64. (In Japanese).

Yamane, T. and Matsuura, K. 1963. Fertilizer management for dry seeded rice cultivation (1). Relation between timing of nitrogen application and its leaching loss. *Chugoku Nogyo Kenkyu (Agric. Res. in Chugoku District)*, No. 27: 61. (In Japanese).

Yanni, Y.G., Rizk, R.Y., Corich, V., Squartini, A., Ninke, K., Philip-Hollingsworth, S., Orgambide, G., de Bruijn, F., Stolzfus, J., Buckley, D., Schmidt, T.M., Mateos, P.F., Ladha, J.K. and Dazzo, F.B. 1997. Natural endophytic association between *Rhizobium leguminosarum* bv. *Trifolii* and rice roots and assessment of its potential to promote rice growth. *Plant and Soil*, 194: 99–114. (Cited from Arihara, 2000).

Yoneyama, T. and Akao S. 1998. Research strategy in solving problems on extending the nitrogen fixation ability to major nonleguminous plants. 2. N_2 fixation in the rhizosphere and in the plant. *Jpn. J. Soil Sci. Plant Nutr.*, 69: 403–409. (In Japanese).

Yoshida, T(akashi), Kai, H. and Harada, T. 1973. Studies on the build-up of readily decomposable organic matter in soil, III. Carbon mineralization and build-up of readily decomposable organic matter as affected by quality and quantity of applied organic residues. *Sci. Bull. Fac. Agric. Kyushu Univ.*, 28: 37–48. (In Japanese).

Yoshida, T(omio) 1971. Soil Microbiology. *IRRI Annual Rep.*, pp. 24–25, 48–50.

Yoshida, T. and Ancajas, R.R. 1971. Nitrogen fixation by bacteria in the root zone of rice. *Soil Sci. Soc. Am. Proc.*, 35: 156–157.

Yoshida, T. and Broadbent, F.E. 1975. Movement of atmospheric nitrogen in rice plants. *Soil Sci.*, 120: 288–291.

Yoshida, T., Roncal, R.A. and Bautista, E.M. 1973. Atmospheric nitrogen fixation by photosynthetic microorganisms in a submerged Philippine soil. *Soil Sci. Plant Nutr.*, 19: 117–123.

Yoshino, T. and Dei, Y. 1977. Prediction of nitrogen release in paddy soils by means of the concept of effective temperature. *J. Central Agric. Expt. Sta., Japan*, 25: 1–62. (In Japanese).

Chapter 9: Fertility Considerations for Paddy Soils (II)—Phosphorus and Other Nutrients

9.1 Phosphorus

Amounts of phosphorus, P, sufficient for growing crops are not usually supplied by irrigation water, nor by any other natural means. Therefore, a soil's P status is primarily determined by the total P content of the parent material. However, even when the total P level is the same, the availability of P can vary greatly from one soil to another. Submergence of paddy fields exerts a great influence on P availability. The P balance in paddy soils is different from that in upland soils and this is examined first.

9.1.1 Phosphorus balance in paddy soils

Figure 9.1, taken from Nanzyo (1996), shows the P balance for one cropping season of rice in paddy soil. The figure was compiled from reports from Omagari, Tohoku District (Sumida *et al.*, 1990) for P content in stalks, leaves, and ears, and from the Kasumigaura area near Tokyo (Takamura *et al.*, 1976; Takamura *et al.*, 1977a,b) for other values of P input and output monitored at three sites. The fertilizer P application rate at Omagari was within the range of those at the Kasumigaura sites. The soils at these monitoring sites included ando soils (Udands) and alluvial soils. The P content in stubble and roots was 0.05–0.32 g P_2O_5 m^{-2} for the Kasumigaura sites.

The figure indicates that fertilizers are the largest P source, followed by stalks and leaves. The largest P output is through the harvest of panicles, drainage being minor. The highest figure of P lost by drainage, 0.44 g m^{-2} or 4.4 kg ha^{-1}, occurred just once due to surface drainage before transplanting (see Figure 8.4).

The amount of P taken up by a crop of rice at the Omagari site was 5.2 mg m^{-2} or 52 kg ha^{-1}. The pattern of P uptake and translocation over time is shown in Figure 9.2. A large amount of P is taken up by rice during the maximum tillering to the heading stage. During this period, soil is kept

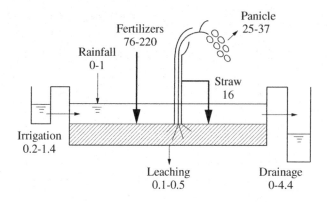

Figure 9.1
Phosphorus balance for one crop of rice (in kg $P_2O_5 ha^{-1}$) (Source: Nanzyo, 1996)

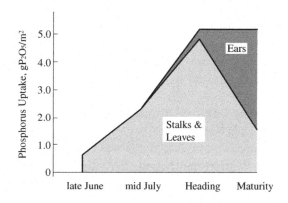

Figure 9.2
Phosphorus uptake pattern of rice and its distribution in the plant
(Source: Nanzyo, 1996)

strongly reduced and this facilitates the dissolution of P by reduction of the Fe-bound fraction. When mid-term drying is practiced, available P decreases for a while, but upon re-submergence it recovers (Shoji and Mae, 1984).

9.1.2 Solubilization of phosphorus in paddy soils
A relative advantage of paddy soils over upland soils with respect to P is clearly shown in Table 8.3 on page 134. This table, previously cited in relation to N, was prepared by Kawasaki (1953) to summarize the results of

Table 9.1   *Phosphorus uptake by rice and wheat and soil phosphorus status in a paddy soil derived from granitic alluvia (kg a⁻¹)*

Plot	Rice brown rice	Rice P uptake	Wheat wheat grain	Wheat P uptake	Annual P uptake	Soil total P g kg⁻¹
Reduced N						
P for each crop	37.0	0.32	24.6	0.29	0.61	1.91
P for rice	34.6	0.29	18.6	0.19	0.48	1.12
P for wheat	34.6	0.30	25.6	0.25	0.55	1.19
No P	33.2	0.25	3.0	0.02	0.27	0.90
Standard N (along with farmyard manure)						
P for each crop	41.3	0.41	29.7	0.30	0.71	1.93
P for rice	43.6	0.39	23.3	0.23	0.62	1.36
P for wheat	43.6	0.39	29.4	0.32	0.71	1.38
No P	40.1	0.35	15.6	0.13	0.48	0.97

Notes:
(1) Figures are the mean for the experimental period, rice 1956–59 and wheat 1955–58. (2) Soil total P includes all forms of mineral and organic P measured right after the experiments. (3) This work was a part of a long-term continuous trial since 1930 at the Hiroshima Agr. Expt. Sta.
(Source: Komoto, 1974)

fertilizer trials, both in pots and in fields, conducted at many different locations throughout Japan. Taking the yield of the complete plot as 100, the table shows indices for the yield of rice, upland rice, and upland cereals, including wheat, barley, naked barley, and two-row barley. No-P plots of rice show indices of 89 for pot trials and 95 for field trials, while the yield of No-P plots of upland rice falls to 79 and 84 for pot and field conditions, respectively. The upland cereals, wheat and barley, are even more susceptible to a shortage of P, giving yield indices of only 62 and 69, for pot and field trials, respectively. The number of experiments suggests that this difference among crop types is significant.

Table 9.1 shows that rice can absorb P even from the No-P plot, but wheat planted as the second crop on the same field absorbs much less P. All of these experimental results indicate that lowland rice is in an advantageous position relative to upland cereals, and this is certainly caused by the practice of submerging of paddy fields.

According to Furukawa (1978), the enhanced availability of P under submerged soil conditions has been attributed to:

- the reduction of ferric phosphate to ferrous phosphate and the concomitant release of phosphate anions;
- the release of occluded phosphate by the reduction of hydrated ferric oxide coatings;

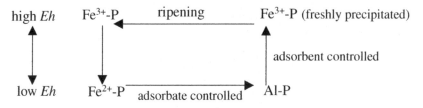

Figure 9.3
Cyclic change of the mineral phosphate forms in the redox process
(Source: Furukawa, 1978)

- the higher solubility of ferric and aluminum phosphates due to pH rise by reduction;
- the dissolution of phosphate from ferric and aluminum phosphates by organic acids;
- the mineralization of organic phosphates;
- the release of phosphate by hydrogen sulfide.

The first two mechanisms are considered to be most significant. Patrick (1964) showed a marked increase in extractable P when the Eh was lowered below 200 mV, the Eh level at which ferric ion also began to be reduced to the ferrous form. Cholitkul and Tyner (1971), using the Chang and Jackson fractionation method, reported large amounts of Fe-bound P and occluded P in paddy soils. Kawaguchi and Kyuma (1969) also showed that many Thai soils contain a high proportion (66–80%) of total fractionated P in these two forms, irrespective of the parent material and age of the soils. In this connection, Furukawa presumes a P transformation in the submergence-drainage cycle of paddy soils, as shown schematically in Figure 9.3.

The reactions in each step of transformation may be formulated as follows:

$$\overset{\text{amorph. Al(OH)}_3}{Fe^{3+}-P \rightarrow Fe^{2+}-P + PO_4^{3-} \rightarrow Al(OH)_3-P + Fe^{2+}}$$
(by reduction)
$$Al(OH)_3 - P + Fe(OH)_3 \text{ (freshly precipitated)} \rightarrow Fe(OH)_3 - P + Al(OH)_3$$
(by oxidation)

Thus, P bound to amorphous hydrated aluminum oxides would be stable temporarily during the submerged period, but upon drainage and oxidation it is transferred to newly precipitated hydrated ferric oxide and undergo ripening as Fe-bound phosphate.

Concerning the reduced form of iron phosphate, a model experiment carried out by Nanzyo (1990) revealed that an amorphous hydrated ferric oxide-P sorption complex produced a ferrous phosphate compound that showed a similar IR spectral pattern to vivianite when reduced by a reductant with (z)-1,2-endiol partial structure, such as ascorbic acid or catechol.

The third mechanism is related to the pH shift of soil under reduced conditions. Figure 9.4 presents a solubility diagram of various phosphate compounds. Iron and aluminum phosphate, as the dominant forms of P in acid soils, increase their solubility as the pH of the medium rises, and that ten times of solubility increase for each unit rise in pH, as shown in the figure for strengite ($FePO_4 \cdot 2H_2O$ or $Fe(OH)_2 \cdot H_2PO_4$) and variscite ($AlPO_4 \cdot 2H_2O$ or $Al(OH)_2 \cdot H_2PO_4$). Even in soils with an alkaline reaction, the lowering of the pH towards neutrality upon reduction should increase the solubility of calcium-bound P, which is the dominant phosphate form in such a soil.

The fourth mechanism is often mentioned in relation to the chelating ability of organic anions, but it is not particularly specific to submerged soil conditions. Of course, in reduced conditions, many kinds of partially decomposed organic compounds that have chelating ability are produced by the fermentation of organic matter.

The fifth mechanism is not necessarily specific to paddy soils. However, in some paddy soils, organic phosphates constitute a high percentage of total P and are found to contribute significantly to an increase in P availability upon submergence. Furukawa and Kawaguchi (1969) reported that iron phytates are particularly important in this mechanism.

The sixth mechanism operating in a strongly reduced condition is expressed by the following equation (Ojima and Kawaguchi, 1959):

$$Fe_3(PO_4)_2 + 3H_2S = 3FeS + 2H_3PO_4$$

The evolution of hydrogen sulfide in relatively iron-poor degraded paddy fields releases some of the Fe-bound P in this mechanism.

Here, some of the mechanisms of phosphate dissolution in the soil are reexamined in the light of theoretical studies on phosphate chemistry.

The enhancement of phosphate solubility by the reduction of ferric phosphate used to be explained by assuming that ferrous phosphate was more soluble than ferric phosphate (Aoki, 1941). This assumption, however, does not hold. The solubility product data of ferric phosphate or strengite ranges from 10^{-23}–10^{-26}, while that of ferrous phosphate or vivianite ranges from 10^{-32}–10^{-36} depending on the source. Of course, the reduction of

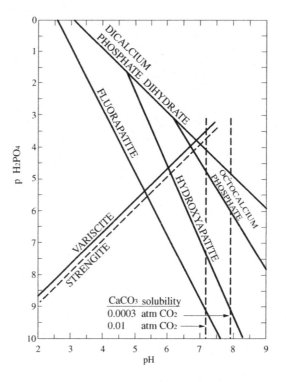

Figure 9.4
Solubility of various phosphates in soil at 25°C and at pCa = 2.50.
(Source: Lindsay and Moreno, 1960)

strengite to vivianite releases some of the phosphate, which may well be utilized by plants. However, Tanaka *et al.* (1969) claimed, that the liberated P in this process would eventually be precipitated as aluminum phosphate or variscite, based on the fitting of their experimental data to the solubility product curves. They concluded that the enhanced solubility of phosphates in paddy soils should be considered to be the result of a rise in pH of the reduced soil system.

Later, Iimura (1982) attempted to explain the higher phosphate solubility in the pH ranges above 7 and below 4–5 under submerged conditions relative to upland conditions. He said that the formation of vivianite in the presence of siderite would explain, at least qualitatively, changes in phosphate solubility.

From the viewpoint of chemical equilibria, Lindsay (1979) showed that at lower redox levels, both strengite and variscite could be transformed to vivianite, regardless of which oxides determine iron solubility. He showed that at a pH of 7 this transformation should occur in a pE range between -3 and $+1.3$ or in an *Eh* range between -177 and $+77$ mV. Lindsay's explanation for the higher phosphate availability to rice, in spite of the strong suppression of phosphate solubility by vivianite in the bulk of the soil, was based on a relatively oxidized rice rhizosphere due to the transmission of oxygen through the aerenchyma of rice stems. Concerning the dominance of iron phosphate in paddy soils, Lindsay (1979) considers that during the oxidation process after drainage of the submergence water, vivianite dissolves and Fe^{2+} is oxidized to Fe^{III}, and phosphate precipitates. Since the amorphous Fe^{III} oxides initially precipitated have higher iron solubility, the solubility of ferric phosphate is depressed below that of variscite, favoring the precipitation of iron rather than aluminum phosphates.

These various mechanisms of P solubilization, only indicate the relative advantage of paddy soils in the P economy compared with upland soils. This does not exclude the possibility of P deficiencies in paddy soils. Table 9.2 illustrates the level of total P in relation to the yield of rice in long-term No-P plots at Hiroshima and Shiga Agricultural Experiment Stations. The data shows that the No-P plot in Shiga is lower in total P than the Hiroshima plot, in spite of the fact that both had almost the same total P at the beginning of the experiment. The yield was around 4 t ha^{-1} for both stations at the beginning, but began to decline when the total P level fell below 400 mg P_2O_5 kg^{-1} soil. The yield at Shiga was as low as 40% of the control plot when total P declined to 250 mg. Also, paddy soil from Battambang, Cambodia containing 440 mg P_2O_5 kg^{-1} soil showed a severe P deficiency in a field trial.

9.1.3 Soil available phosphorus and rice performance

Shiga (1973) compared various methods commonly used for determining the available P by correlating the amount of P in the soil and that taken up by rice. He found that the 2.5% acetic acid extraction method and the Bray-Kurz No. 2 method were the most satisfactory for paddy soils. He also showed that P uptake by rice has a higher correlation with available P measured in soil after submerged incubation than with that measured in air dried soil. Shiga and Yamaguchi (1977) considered that some 20–50% of available P measured after submerged incubation would become available in the field condition.

In spite of this finding, most of the data for available P has so far been obtained with air-dried soils because submerged incubation takes time.

Table 9.2 Soil total phosphorus and rice yield in alluvial paddy soils

Location	Year of sampling	Soil total P$_2$O$_5$, mg kg^{-1}		Mean brown rice yield for 3 yrs, kg a^{-1}		
		P plot	No-P plot	P plot	No-P plot	No-P/P %
Shiga Agr.	1937	810	320	38.5	30.8	80
Expt. Sta.	1956	1646	242	42.7	16.5	39
Hiroshima Agr.	1936	-	490	37.4	39.2	105
Expt. Sta.	1940	1010	400	39.0	38.9	100
	1942	1040	380	38.3	37.4	98
	1963	1380	340	48.2	44.1	91

Note2:
(1) Both stations started the trials in 1930. The soil total P$_2$O$_5$ at the starting year was 689 mg kg^{-1} at Shiga and 690 mg kg^{-1} at Hiroshima.
(2) The soil at Shiga is derived from granite, slate and sandstone, while that at Hiroshima is derived from granite.
(Source: Shiga Agr. Expt. Sta., 1963; Hiroshima Agr. Expt. Sta., Annual Reports, 1968–1970)

Recently, as a measure to replace submerged incubation, overnight (16 hours) treatment with 1% ascorbic acid as a reductant was proposed by Nanzyo *et al.* (1996). A high correlation was found across a wide range of soils between the estimates thus obtained and those obtained using the modified Bray No.2 extraction method after submerged incubation.

Figures 9.5 and 9.6 show the relationship between Bray-P (as measured with air-dried soils) and the performance of rice. In the cool climatic region (Figure 9.5), 300–500 mg P$_2$O$_5$ kg^{-1} soil of Bray-P appears to be necessary to secure a sufficient number of tillers, while in the warm climatic region (Figure 9.6) the corresponding figure would be 150–200 mg.

A relationship between rice yield and the Bray-P level in the soil was also established. Shiga (1976) confirmed that a Bray-P content of 220 mg P$_2$O$_5$ kg^{-1} dry soil was required to attain a normal yield of rice (4–6 t ha^{-1}) in the cool climatic region, while rice yield did not respond when soil Bray-P level was higher than 300 mg P$_2$O$_5$ kg^{-1}. In the warm climatic region, a Bray-P level of 60–100 mg P$_2$O$_5$ kg^{-1} dry soil was the threshold above which no yield response was expected (Komoto, 1971) (see Table 9.3 and Figure 9.7).

The relationship between the performance of rice and soil available P is summarized in Table 9.4. This table shows that the upper limit of soil available P needed to increase the tiller number and yield is higher in the cool climatic region than the warm region. In other words, rice in the cool region requires a higher level of soil available phosphorus than rice in the warm region.

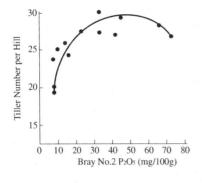

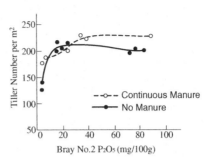

Figure 9.5

Relationship between soil available phosphate and tiller number of rice in volcanogenous paddy soil (Source: Hokkaido Agr. Expt. Sta., 1967)

Figure 9.6

Relationship between soil available phosphate and tiller number at the tillering stage in granitic alluvial paddy soil (Source: Hiroshima Agr. Expt. Sta., 1968–70)

The fate of P applied as fertilizer is a matter of great concern both economically and environmentally. The recovery rate of applied P in paddy soils, as estimated by the subtraction method for a long-term trial, ranges from 10–38% (Sumida *et al.*, 1990) while it is 13–27% (Uwasawa *et al.*, 1980) when estimated by the tracer technique for a thirty-nine day pot experiment. Most of the remaining applied P is temporarily or permanently fixed by soil materials, primarily by sesquioxides in acid soils. As shown below, the leaching loss of P is normally negligible.

Table 9.3 Relationship between soil phosphorus content and rice yield

Plot No.	Soil total P_2O_5 mg kg^{-1}	Bray No.2 P_2O_5 mg kg^{-1}	Brown rice yield index
1	380	36	100
2	400	43	123
3	430	45	122
4	480	68	134
5	570	92	132

Notes: Soil total phosphorus was determined with air-dried soil, while Bray No.2 method was applied to wet soil sampled at tillering stage.
(Source: Komoto, 1971)

(a) At Sapporo, Hokkaido (Northern Japan)

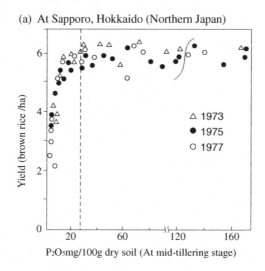

(b) At Higashi-Hiroshima, Hiroshima Prefecture
(Southern Japan)

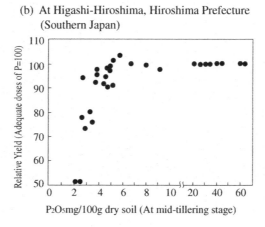

Figure 9.7
Relationship between Bray No. 2 phosphorus and yield of rice at two locations in
Japan (Source: Kogano, 1984; Komoto, 1964; cited from Suzuki, 1997)

9.2 Basic cations

Compared with upland soils, a specific feature of paddy soils is the inflow
of basic cations with irrigation water or floodwater. Thus, in the long run,
the quality and quantity of irrigation water relative to percolation water
determine the base status of paddy soils. Depending on both the balance

Table 9.4   The upper limits of phosphate status of soil and rice plant at which rice response to phosphate fertilizer application may be expected

Region	Mean temp. at tillering stage	For increase in tiller number		For increase in yield	
		Bray No.2 P_2O_5 in soil, mg kg^{-1}	P_2O_5 contents in stalks and leaves at tillering	Bray No.2 P_2O_5 in soil, mg kg^{-1}	P_2O_5 contents in stalks and leaves at tillering
Cool zone	<19°C	350–500	0.8%	200	0.45–0.60%
Warm zone	>20°C	150–200	0.7%	50	0.30–0.45%

(Source: Nat. Inst. Agr. Sci., 1969)

between the inflow and outflow of cations and the nature of the soil material, some paddy soils would be enriched with cations and some would be impoverished.

It is also relevant here to consider the plant uptake of these cations. If the supply by irrigation water exceeds the plant uptake, the level of these basic cations as soil reserves has little to do with the performance of the rice plant. Table 9.5 provides some examples of the amounts taken up by plants. These data compare the amount of different elements taken up by different varieties of rice at different yields.

If it is assumed that the water used for a single crop of rice is 1000 mm, and that the water contains 2 mg kg^{-1} of K, the total supply of K_2O from the water would be about 24 kg ha^{-1}, which is sufficiently large to be considered when fertilizers are to be applied. In a study conducted in Japan by Yoshida (1961), it was shown that the amount of calcium and magnesium supplied by irrigation water was 2 to 6 times the requirement of rice, whereas the supply of silica, potash, P, and N was only 30%, 17%, 1% and 8% of the requirement for the respective elements. These figures vary widely from one place to another, but they certainly show the importance of irrigation water as a source of basic cations and silica.

Table 9.5 Amount of nutrient uptake by rice

Brown rice yield, kg a^{-1}	Cultivar	Amount of nutrient element required to produce 100kg of brown rice, kg								
		N	P	K	Si	Ca	Mg	Mn	Fe	Na
102.4	Norin29	1.9	0.43	2.7	9.4	0.50	-	0.077	-	-
101.5	Kinmaze	2.0	0.32	2.5	10.3	0.40	0.23	-	-	-
75.6	Fujiminori	2.0	0.40	2.3	7.3	0.27	0.22	0.040	0.049	0.059
64.8	Manryo	2.1	0.47	2.3	10.1	0.40	0.27	0.056	0.082	0.083
59.9	Manryo	2.1	0.54	2.0	9.6	0.46	0.27	0.055	0.113	0.224

(Source: Seino et al., 1970)

The losses of these basic cations from paddy soils by leaching have been studied mainly in lysimeter experiments. One of the earliest experiments, conducted by Iwata (1928), showed that losses due to leaching amounted to 30 kg of K_2O, 330 kg of CaO, and 160 kg of MgO per hectare per year. If it is assumed that the total amount of water percolated through a paddy field was 1000 mm (10 mm per day for 100 days), the mean concentration of the cations was 2.5 mg kg^{-1} K, 23.6 mg kg^{-1} Ca and 9.6 mg kg^{-1} Mg, all of which exceed the mean concentration of Japanese river water (see Table 11.1), signifying a tendency toward the impoverishment of the base status of the soil.

Table 9.6 provides data comparing the concentrations of various elements in irrigation water and in drainage water from tiles. Here also, the general tendency of the impoverishment of bases is seen except for one case of Ca.

The effect of soil reduction on the increase of cation concentration in the soil solution should be noted. An increase in the concentration of ferrous ions in the soil solution of submerged soils, causes an increase in all the cation concentrations due to the exchange for Fe^{2+}. This is one of the reasons why the percolation water has high concentrations of bases and why paddy soils tend to be acidified (see Chapter 3 and 7).

9.3 Silica

It has not yet been proven that silica is essential for plant growth, but it has long been known that sufficient quantities of silica in the leaves and stalks of rice beneficially affects the plant by (Suzuki, 1997):

- assisting leaf-blades to stand upright, so reducing mutual shading, and contributing to efficient photosynthesis;

Table 9.6 Elemental composition of irrigation water and tile-drained water (mg kg^{-1})

Field condition	Year	Irrigation water						Tile drainage water					
		NH_4-N	Fe_2O_3	CaO	MgO	K_2O	SiO_2	NH_4-N	Fe_2O_3	CaO	MgO	K_2O	SiO_2
Shonai Substation													
CL to 48 cm deep	1961	0.45	1.5	11.2	2.4	0.52	6.7	2.5	133	18.4	4.3	2.6	21.5
gleyed subsurface	1962	0.27	1.1	4.1	3.6	0.75	7.0	2.4	170	8.2	3.6	3.1	21.2
and subsoil	1965	0.08	0.71	9.6	7.0	1.2	-	2.3	52.4	23.6	11.2	1.9	-
Nakano-sone Field													
LiC to 70cm deep	1961	0.53	4.3	17.6	8.2	1.8	17.6	0.69	51.8	12.9	17.0	1.9	20.4
gleyed subsurface	1962	0.5	3.4	17.8	7.8	2.2	17.8	0.77	100	12.9	8.7	2.2	19.5
and subsoil	1965	0.79	11.4	11.1	4.0	1.8	-	1.4	13.5	11.3	4.2	1.9	-

(Source: Shoji *et al.*, 1972)

- increasing resistance to diseases such as rice blast by forming cuticle-silica layers that may function as barriers to microbial infestation;
- increasing the physical hardness of straw, which may reinforce the tolerance to lodging.

Such beneficial functions have been noted more clearly in years with adverse meteorological conditions, such as low solar radiation and low temperatures.

In Japan, silicate materials have been applied to paddy soils since the 1950s, first as an amendment to *Akiochi*-inclined degraded paddy soils (see Section 9.8). According to Sumida (1996), in 1968, a total of 1.3 Mt of silicate fertilizers were used, but their use gradually declined to just one half that amount, 0.65 Mt, by the mid-1980s, and to a little less than 0.5 Mt in the early 1990s. Silicate material is used more in southwestern Japan (Tokai, Kinki and Shikoku districts) and less in northeastern Japan (Hokkaido, Tohoku, and Kanto districts). This regional difference in silicate application appears to reflect the nature of the soil's parent material. Generally, soils in northeastern Japan are rich in volcanogenous material that contains readily-weatherable silicates, whereas soils in southwestern Japan, except those in Kyushu, are poor in material of volcanic origin.

9.3.1 Assessment of available silica

Silica is one of the major components of soil, and normally comprises as much as 60% or more of the total mineral constituents. Nevertheless, the addition of readily soluble silicate, such as calcium silicate, assists rice growth in many paddy fields because only a proportion of soil silica is available to rice. Kawaguchi and Kyuma (1977) found that total silica is highly negatively correlated with available silica—the higher the total silica content, the lower the available silica.

Imaizumi and Yoshida (1958) studied the nature of available silica in the soil and found that only monomeric and oligomeric orthosilicic acids could be assimilated by rice. They showed that a dilute acid dissolves orthosilicic acid from soil silicates, but it also promotes the polymerization of once dissolved orthosilicic acid to colloidal silica. Therefore, there is an optimum acid concentration at which the maximum dissolution of silicates can be achieved. They also showed that orthosilicic acid and its oligomers, which are soluble in dilute acids, are present in the soil in amorphous combinations with iron and aluminum hydrated oxides.

Imaizumi and Yoshida (1958) developed a pH 4-N acetate buffer extraction method for assessing available silica in the soil and correlated the determined available silica level with the plant silica content and yield response. As a result, they proposed the following threshold values of soil and plant silica content for judging the effectiveness of silica application for rice cultivation in Japan.

Class	SiO_2 in air dried straw	SiO_2 in soil	Judgment
I	< 11%	10.5 mg 100g^{-1}	clear positive response
II	11–13%	10.5–13	positive response
III	> 13%	> 13	no response or even harmful

More recently however, the acetate buffer extraction method has been shown to analyze not only effective natural silicate, but also ineffective residual silicate remaining from previously applied silicate materials. Thus the result does not necessarily reveal the exact response of rice to silicate application (Takahashi and Nonaka, 1986).

Some novel methods were proposed to replace the pH 4 acetate buffer extraction method, such as the submerged incubation method (Takahashi and Nonaka, 1986) that simulates paddy soil conditions, and the successive supernatant method (Kitada *et al.*, 1992a) that collects supernatants from successively repeated submerged incubations to simulate the uptake of dissolved silica by rice.

Sumida (1991) proposed the concept of 'readily dissoluble silica' based on his method of evaluating the silica dissolution/adsorption characteristics of soil. He found that when soil is incubated at 30°C for five days with the addition of silicate solution of varying concentrations (0–100 mg SiO_2 litre^{-1} 0.01 M $CaCl_2$), a change in the silica concentration Y from the added silica concentration X could be expressed by the following linear regression equation:

$$X/a + Y/b = 1, \text{ or } Y = b - (b/a)X \qquad a, b > 0$$

Here, b is the silica concentration of an extract with no addition of silica, and a is the silica concentration of a solution, the addition of which causes no change in silica concentration after incubation, thus corresponding to the 'dissolution/adsorption threshold point (DATP)'. The slope b/a may be used as an index for the ease of bringing the silica concentration of an extract to that of the DATP.

In practice, the silica concentration of the extract at the end of incubation, U (mg litre^{-1}), is plotted against the amount of silica dissolution or adsorption, V (mg 100g^{-1}), where:

$$U = X + Y$$
$$V = Y/10r$$

Here, 'r' is the ratio of soil to silicate solution, ranging from 1:10 to 1:100. The plot is expressed by:

$$U/a + V/c = 1, \text{ or } V = c - (c/a)\,U \qquad a, c > 0$$

where:

$$c = a \cdot (b/a)/\{10r(1 - b/a)\} = a \cdot b/\{10r(a - b)\}$$

The 'c' is the extrapolated V intercept of the dissolution/adsorption regression line and is related to the DATP or a, and also to b/a. The value a expresses the static dissolved silica concentration when there is no rice uptake, and b/a signifies the capacity to replenish the silica concentration of the soil solution when the uptake by rice has decreased its concentration. The value 'c' expresses the amount of silica in an extract when a very large volume of water is used to make its silica concentration infinitesimally low. Thus, it is considered to correspond to the total amount of water-soluble silica under the given incubation conditions and, is called 'readily dissoluble silica (RDS)'. RDS has a higher correlation with the silica uptake of rice compared with the available silica measured by the pH 4 acetate buffer extraction method.

Using many measurements of silica content in rice and soils in Tohoku District, Sumida (1992) proposed the criteria shown in Table 9.7 to direct silicate fertilizer application in cool climatic regions, where silica gives rice 'cold-tolerance', particularly through resistance to blast disease.

Although the concept of RDS is sound and acceptable, the method of its determination is rather too involved to use in a routine assessment of available soil silica. Recently, two phosphate buffer extraction methods were proposed, in which phosphate replaces adsorbed silicate. They are based on the assumption that silicic acid taken up by rice from paddy soils mainly originates from the fraction adsorbed on the soil solid phase. Shigezumi *et al.* (2002) used a 20 mM phosphate buffer (pH 6.95) solution with a shorter incubation time (5 h at 40°C), whereas Kato (1998) adopted a 40 mM phosphate buffer (pH 6.2) with a longer incubation time (24 h at 40°C). The latter method gives a higher estimate of available silica that is comparable with Sumida's RDS. For some soils, such as ando soils, neither

Table 9.7   Criteria for application of Ca silicate for paddy soils in the cool climatic zone

Readily dissoluble silica content in paddy soil	Silica content in rice plant at heading	Ca silicate application strategy[1]
<300 mg kg[-1]	<70 mg g[-1]	annual[2]
300–400 mg kg[-1]	70–80 mg g[-1]	reduced[3]
>400 mg kg[-1]	>80 mg g[-1]	none

Notes:
(1) As Ca silicate application accelerates reduction of soil organic nitrogen, the generally recommended amount of organic matter should be applied.
(2) Rate of application should be around 2000 kg ha[-1].
(3) Preferably apply every other year with proper testings.
(Source: Sumida, 1992)

method produces a good correlation. However, silica extracted using a phosphate buffer solution generally correlates better with silica taken up by rice than that extracted with the pH 4 acetate buffer solution. As a result, the phosphate buffer pH 6.2 extraction method and the submerged incubation method (Takahashi and Nonaka, 1986), were recently adopted as routine analytical procedures for available soil silica in the national soil-testing project of the Japanese Ministry of Agriculture, so replacing the acetate buffer method.

9.3.2 Use of silicate materials as amendments

Silicate basic slag from steel mills is most commonly used for amending available silica in paddy soil. Its major constituent is either Ca-metasilicate ($CaO \cdot SiO_3$) or Ca-orthosilicate ($2CaO \cdot SiO_2$). It also contains small amounts of Mg, Fe, Mn, and B. The Ca-metasilicate is decomposed in the soil, reacting with carbonic acid. Straw and farmyard manure can also be used as amendments.

Takahashi (1981) reported that the fertilizer effect of basic slag is controlled by its composition. When its molar ratio $(CaO + MgO)/SiO_2$ is lower than 1, as with acidic slag, the fertilizer effect is low. The crystallinity of slag impacts on its fertilizer effect, which is higher when slag is slowly cooled and well crystallized. Also, the finer the grain size, the higher the fertilizer effect. However, the official method of assessing fertilizer effects in Japan, based on 0.5 M HCl soluble silica, does not necessarily reflect these differences.

Sumida (1991) confirmed that slag silicate material contains two fractions of silica, one being the fraction that is made available to rice at the time of heading in the year of application and the other is the fraction that provides a certain residual effect in successive years. Application of organic material, such as rice straw and straw manure, not only supplies silica in the material, but also promotes soil reduction that enhances solubilization and the supply of soil silica.

The recovery rate of applied silica by rice is reported to be 20–50% for slag silicate, as measured by the subtraction method. Sumida (1991) gave figures of about 30% for slag silicate, 6% for rice straw and 3% for rice straw manure. Kato and Owa (1990) reported that the [30]Si tracer method tends to give a higher recovery rate than the subtraction method, particularly during the earlier growth stages before heading. The recovery rate of slag silicate, as measured by the use of the [28]Si tracer method, was 32–33%. They shared with Sumida the view that silica uptake by rice during the period after heading was insignificant.

Available silica is leached relatively rapidly out of the soil. Table 9.8 shows an example of the leaching process, using soils of the Kojima polder. The decline in soluble silica is quite remarkable, and the rice yield falls with it.

Four other factors should be considered as regards the available silica status in paddy soils. One is the supply of silica by irrigation water. As stated earlier, in one instance in Japan about 30% of the crop requirement was supplied by irrigation water. In tropical countries, the silica content of river

Table 9.8 Leaching loss of silica from the surface soil of a dry paddy field

Soil properties and performance of rice	Years of rice cultivation after polder land reclamation					
	0	2	60	140	240	1,000
pH (H$_2$O)	6.30	6.12	5.12	5.04	5.02	5.28
Clay %	19.13	19.98	25.84	24.24	17.12	6.75
Hot conc. HCl-soluble SiO$_2$ %	24.1	21.4	20.8	18.9	14.8	11.1
0.5N HCL-soluble SiO$_2$ %	0.79	0.58	0.47	0.47	0.32	0.33
2% Na$_2$CO$_3$-soluble SiO$_2$ %	0.85	0.69	0.60	0.51	0.49	0.36
Pot expt. with rice unhusked grains, g		73	64	61	53	54
straw weight, g		104	112	98	81	68
SiO$_2$ in straw, %		15.4	13.6	11.5	10.7	11.9
Regional mean brown rice yield, kg 10a^{-1}		450.0	385.5	364.5	355.5	–

(Source: Kubota, 1961)

water is normally high, reflecting a higher rate of weathering of soil materials. It is also high in areas of active volcanism.

The second factor is the reduction of paddy soils during the cropping season. As stated above, some of the soluble silica is stabilized in the soil as amorphous iron silicates, which undergo reduction to release available silica for plant uptake.

The third factor is the influence of land use patterns on soil silica availability. Recently, various patterns of paddy/upland crop rotation have been practiced in Japan. Using an index devised to reflect the specific physical nature of upland soils relative to paddy soils, Kitada *et al.* (1992b) showed that silica taken up by rice was invariably higher when the index pointed to the more distinct upland nature of a soil. Sumida and Kato (2001) also demonstrated that both the silica extracted with the phosphate buffer (pH6.2) method and that taken up by rice increased dramatically in the first year of conversion from uplands to paddy fields.

The fourth factor is relevant to the no-tillage or no-puddling practice that is adopted to reduce both labor and wet injury caused by excessive soil moisture for upland crops in a paddy/upland rotation. Kato *et al.* (2001) showed that these soil management practices tend to increase the available silica in paddy soils.

9.4 Iron

9.4.1 Iron-deficient paddy soils or degraded paddy soils

Iron is one of the most noteworthy elements in paddy soils because it is abundant and undergoes redox transformation. Here the roles of iron in its relation to the fertility of paddy soils are discussed.

The term 'iron-deficient soil' is often used in Japan. This soil is generally sandy in texture and light in color. The rice plants grown in this type of soil are likely to suffer from the '*akiochi* effect' (see 9.8) and are low yielding. This is not due to a physiological iron deficiency, but is often caused by the toxic effect of hydrogen sulfide (see 9.6) under a strongly reductive soil conditions. The threshold iron content for iron-deficient paddy soils in Japan is $1\% \ Fe_2O_3$, as determined by the Mg-ribbon reduction method (Kawaguchi and Matsuo, 1955).

As stated earlier, iron seems to be the most probable potential-determining element of reduced soil and therefore its low content in soil relative to the content of readily decomposable organic matter causes a strong reduction. Hydrogen sulfide and methane are vigorously emitted together with volatile fatty acids that are also toxic to rice plants. Moreover, due to the paucity of

Table 9.9 Comparison between degraded and normal paddy soils

Soil	Horizon thickness cm	Free Fe$_2$O$_3$ %	Hot HCl Fe$_2$O$_3$ %	Hot HCl P$_2$O$_5$ %	TN %	TC %	NH$_4$-N mg 100g^{-1}	H$_2$S emission
Strongly degraded paddy soil								
Plow layer	0–12	0.14	0.69	0.075	0.22	1.97	9.0	yes
Bleached plowsole	12–16	0.19	0.79	0.052	0.08	0.48	4.5	yes
Iron illuv. horizon	16–20	1.17	3.01	0.142	0.08	0.44	3.9	no
Normal paddy soil								
Plow layer, upper	0–10	0.21	1.13	0.128	0.24	2.47	9	-
Plow layer, lower	10–14	0.3	1.17	0.103	0.23	2.17	6.9	-
Iron illuv. plowsole	14–18	1.54	3.06	0.080	0.08	0.47	2.4	-

(Source: Shioiri, cited from Kawaguchi (Ed.) *Paddy Soil Science*, 1978)

iron, a protective iron coating on rice roots is not formed, so rice roots are easily damaged and the normal growth of rice is strongly inhibited even when other nutrient elements are supplied in sufficient quantity. A typical example of iron-deficient soil is given in Table 9.9. Such iron-deficient soils are also called 'degraded' paddy soils.

The remedy for such a soil is incorporation of iron-rich material. It has been shown experimentally that the addition of ferric hydroxide sustains the redox potential of the medium at a relatively high level. In practice, this is done by soil dressing—applying red-colored (iron-rich) mountain soil onto iron-deficient soils. The quantity of iron-rich soil to be applied and the effect of soil dressing on the productivity of iron-deficient degraded soils are illustrated in Figure 9.8.

As discussed later in relation to *akiochi*, the effect of soil dressing is not solely due to the iron, but also due to the addition of better quality clay and other nutrients, particularly silica, Ca, Mg, K, and Mn.

One special technique for dressing soil is a type of *'colmatage'* irrigation, which is irrigation with clay-suspended muddy water. This method allows a large area of degraded soil to be treated relatively easily.

9.4.2 Iron toxicity problems

Another fertility problem related to iron is iron toxicity. As the soil is reduced, the ferrous iron concentration in the soil solution increases. At times, it can exceed 500 mg litre^{-1}. However, according to Takagi (1958), this high level of ferrous iron does not itself necessarily cause iron toxicity.

Iron toxicity seems to occur in two groups of soils—acid sulfate soils and strongly depleted soils of Oxisol nature. The former is widely

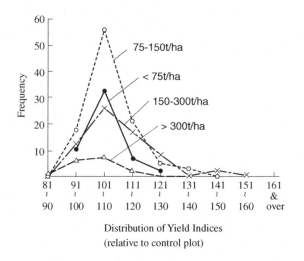

Figure 9.8
Effect of iron-rich red soil materials on degraded paddy soils
(Source: Min. Agr., cited from Kawacuchi (Ed.) Paddy Soil Science, *1978)*

distributed in Southeast Asia, but the latter is rather limited in paddy soils. 'Bronzing', which is thought to be a symptom of iron toxicity, occurs in Sri Lanka in a small valley filled with sandy sediments from surrounding low hills that are covered with Oxisol-type soils. Tanaka and Yoshida (1970) observed typical iron toxicity symptoms on sandy lateritic soils at Bhubaneshwar, Orissa State in India, which is an area with many abandoned laterite quarries.

According to Tanaka and Yoshida (1970), the symptoms of iron toxicity are not directly related to a high iron content either in the medium or in the plant. They seem to think that it is a result of interaction between a high iron content in the soil solution and a deficiency of one or more of such essential nutrients as potassium and phosphorus. According to Tadano (1975), rice plants have a built-in protective mechanism that enables them to exclude excessive amounts of ferrous ions, if they are nutritionally healthy.

In the case of Oxisol-type soils, such as those observed in Sri Lanka and India, the status of all nutrients is usually low. In addition, the active iron content is also often very low. Under such conditions, hydrogen sulfide from the reduction process inhibits the respiratory function of plants. Since the protective mechanism of the rice plant depends on respiration, sulfide injury can fatally affect the mechanism and allow excessive iron uptake, which causes iron toxicity even with relatively low concentrations of ferrous iron in the soil solution. Yamada (1959) and Inada (1966) argued in their studies on bronzing in Sri Lanka that root damage by hydrogen sulfide is a possible primary cause of iron toxicity. Van Breemen and Moormann (1978) studied bronzing-affected areas in Sri Lanka and suggested that bronzing could have been caused by interflow water high in dissolved iron.

The situation may be different in acid sulfate soils. First of all, an extremely low pH due to free sulfuric acid produced by pyrite oxidation would weaken the mechanism in rice roots that protects them from taking up excessive amounts of iron. Also, the poor nutritional status of a plant aggravates the toxicity. Thus, a generally low P level in acid sulfate soils is a major concern. High salinity due to NaCl or $MgCl_2$ is known to decrease the oxidizing power of rice roots, so enhancing iron toxicity (Tadano, 1975). Iron toxicity partly due to this cause, could be widespread in many young acid sulfate soils.

Although iron toxicity has received much attention, further studies are needed to be able to effectively counteract local incidences of this disorder.

9.5 Manganese

Manganese is another element that undergoes redox transformation. However, its occurence in soil is normally 1/10 or less that of iron, so, is not considered important in determining the soil Eh.

Manganese is an essential element for plants. Therefore, its physiological deficiencies and toxicities have been a primary concern of soil scientists. The solubility of manganese in upland soils is mainly determined by the soil reaction. For alkaline reactions above pH 8.5, its solubility is critically low, while for acid reactions below pH4.5, its solubility is sufficiently high . The redox condition affects the solubility of manganese more significantly in the range between the two extremes.

As the Eh at which Mn^{4+} becomes unstable is about 300 mV (Patrick and Reddy, 1978), manganese compounds in the soil are reduced relatively easily to the more soluble manganous form. In this way, manganese is almost as mobile as other basic cations in paddy soils. This implies that there are many paddy soils from which manganese has been leached. The total area of manganese-deficient soils in Japan is reported to be 149,000 ha.

In Japan, the threshold value for manganese deficiency for paddy soils is set empirically at 25 mg MnO kg^{-1} air-dried soil, which is extractable with a neutral N NH$_4$-acetate solution containing 0.2% hydroquinone (readily reducible manganese). Soils containing manganese below this level are expected to respond positively to manganese, which is usually applied as manganese sulfate or manganese containing silicate slag at the rate of 20kg MnO ha^{-1}. However, as manganese deficiency is often part of the general *Akiochi* phenomenon of rice that is grown on degraded soils, manganese application alone is rarely sufficient.

Manganese toxicity occurs in soil with readily reducible manganese of more than 300 mg kg^{-1}. Such soil often occurs in those lowlands that have ground-water containing a high amount of manganese. Toxicity symptoms are: retarded initial growth, a lower number of tillers, and chlorosis when toxicity is severe.

According to Sumida (personal communication), in the paddy/upland rotational system that become more common recently, manganese toxicity sometimes occurs for upland crops, such as soybeans, in the first year of conversion from paddy to upland. This may be caused by the slow oxidation of the soil matrix or by acidification of soil upon oxidation. Correction of the pH to 5.5–6 appears to be effective as a counter-measure.

9.6 Sulfur

Sulfur is an essential element for plants, but a physiological deficiency is rarely a problem in rice cultivation. Nevertheless, sulfur is an important element in paddy soils, as it undergoes redox transformation and produces substances toxic to rice under reduced conditions.

In an oxidized soil, sulfate is the stable form and in this form it is utilized by the plant. However, when the soil is strongly reduced, it is reduced to sulfide form. Patrick and Reddy (1978) gave the range of redox potentials at which sulfate becomes unstable and transforms to sulfide (−120 ~ −180 mV). Therefore, sulfide appears only at a later stage in the sequential reduction process.

Osugi and Kawaguchi (1938) confirmed sulfate reduction in paddy soils, and related it to the physiological disorder of rice known as 'root rot'. They explained that sulfate reduction occurs most vigorously at 35°C and at pH 6–8, and that it is accelerated by the addition of fresh organic matter. However, disorders of rice due to sulfate reduction are only seen in iron-deficient soils. When the free iron oxide content is high, even though sulfate reduction occurs, free hydrogen sulfide is not produced and no root rot occurs. As a practical counter-measure to the disorder, the incorporation of iron-rich material, as explained above, may be the most effective approach.

Shiga and Suzuki (1965) found that in mineral soils, free hydrogen sulfide was not detected in the soil solution when the active iron oxide content measured with the $(NH_4)_2S$ extraction method was > 0.2%. They found that the active iron content as measured by the Truog method did not correlate as well with the evolution of free hydrogen sulfide as did that measured by the $(NH_4)_2S$ method.

The chemistry of sulfide compounds in paddy soils is discussed next using available physico-chemical data. Hydrogen sulfide, H_2S, is dissociated in two steps:

$$H_2S = HS^- + H^+ \qquad \log K_1 = -7.0 \qquad (1)$$
$$HS^- = S^{2-} + H^+ \qquad \log K_2 = -13.0 \qquad (2)$$

When the total dissolved sulfide, S_T, is expressed as:

$$S_T = H_2S(aq) + HS^- + S^{2-} \qquad (3)$$

the ionization fraction for each sulfide species is derived as follows (see Chapter 6):

$$[H_2S] = S_T [H^+]^2 / ([H^+]^2 + K_1[H^+] + K_1K_2) \qquad (4)$$
$$[HS^-] = S_T K_1 [H^+] / ([H^+]^2 + K_1[H^+] + K_1K_2) \qquad (5)$$
$$[S^{2-}] = S_T K_1 K_2 / ([H^+]^2 + K_1[H^+] + K_1K_2) \qquad (6)$$

The solubility of S^{2-} is regulated by Fe^{2+}, through the solubility product of FeS (amorphous):

$$K_{sp} = [Fe^{2+}][S^{2-}] = 10^{-16.9} \qquad (7)$$

whereby Fe^{2+} solubility is also controlled by the $FeCO_3$-CO_2-H_2O system (see Chapter 6) in strongly reduced soil.

$$K_{sp} = [Fe^{2+}][CO_3^{2-}] = 10^{-10.7} \qquad (8)$$

Thus, it is appropriate to combine (7) and (8) and consider the pH range around neutrality, where HCO_3^- is dominant:

$$\begin{array}{lr} & \log K \\ \text{FeS(s)} = \text{Fe}^{2+} + \text{S}^{2-} & -16.9 \\ \text{Fe}^{2+} + \text{CO}_3^{2-} = \text{FeCO}_3(\text{s}) & 10.7 \\ \text{HCO}_3^- = \text{H}^+ + \text{CO}_3^{2-} & -10.3 \\ \hline \text{FeS(s)} + \text{HCO}_3^- = \text{FeCO}_3(\text{s}) + \text{H}^+ + \text{S}^{2-} & -16.5 \end{array}$$

$$\log[\text{S}^{2-}] = -16.5 + \log[\text{HCO}_3^-] + \text{pH} \tag{9}$$

When the medium pH is near neutral, from (6):

$$[\text{S}^{2-}] = S_T K_1 K_2 / [\text{H}^+]^2$$

and so

$$\log[\text{S}^{2-}] = \log S_T + \log K_1 + \log K_2 + 2\text{pH} \tag{10}$$

When (9) is equated with (10):

$$\log S_T = 3.5 + \log[\text{HCO}_3^-] - \text{pH} \tag{11}$$

Assuming the total dissolved carbonate, C_T, to be 10^{-2} M, the total sulfide concentration for the pH range of 6.5–7.0 is calculated as:

$$\log S_T = -5.0 \sim -5.5 \tag{12}$$

The total sulfide solubility of $10^{-5.5}$ M is almost exactly equivalent to 0.1 mg litre^{-1} of sulfur. If sulfate reduction proceeds vigorously and the sulfide formed is not removed by precipitating FeS, the amount beyond this limit is emitted as hydrogen sulfide gas. Under acid conditions, the total sulfide solubility decreases and the share of undissociated hydrogen sulfide increases, so that the latter is more readily emitted.

The presence of readily reducible iron oxides guarantees the supply of Fe^{2+} ions to maintain the solubility of sulfides at the level predicted above, so preventing the evolution of hydrogen sulfide gas. Iron also plays another important role, as a large amount of free or active iron oxides assures a pH rise under submergence. Takai et al. (1957 a, b) used soils from Nagano and Sanage to investigate this (see Figure 3.4). The Nagano soil contained 1.10% of free iron oxide, while the Sanage soil had only 0.09%. The pH of the Nagano soil rose above 6.4, whereas the pH of the Sanage soil stayed below 6, even though the Eh fell in both cases. This effect of elevating the alkalinity of the soil solution by the dissolution of ferrous ions is critically important for suppressing the toxic effect of hydrogen sulfide.

The toxic effects of sulfide comes from its inhibition of the respiratory enzyme system. Therefore, it affects the positive metabolic uptake of various nutrients, as seen in Figure 9.9. The inhibitory effect of H_2S on nutrient uptake increases in the following order: $P > K > Si > NH_4\text{-}N > Mn > H_2O > MgO > CaO$.

Mitsui *et al.* (1953) numerically expressed this relationship, using the 'inhibition coefficient of nutrient uptake', or IC, which is written as follows:

$$IC = (A - B)/A \times 100$$

where A is 'nutrient uptake in the control plot' and B is 'nutrient uptake in the treatment plot'. Application of this formula produced the following numerical values for each nutrient with respect to hydrogen sulfide: K_2O (182), P_2O_5 (147) > SiO_2 (107), SO_3 (100), Br (85) > MnO (43), $NH_4\text{-}N$ (42), H_2O (46) > MgO (24), CaO (16). Although the order of nutrients is slightly different from that given above, it is interesting to note that the most strongly inhibited elements are the ones known to be deficient in the degraded or *akiochi* soils. This again confirms that the *akiochi* phenomenon is related to the toxic effect exerted by hydrogen sulfide.

Sulfur deficiency is known to occur normally in inland regions distant from the ocean. It was suspected in the 'yellow leaf disease' occurring in

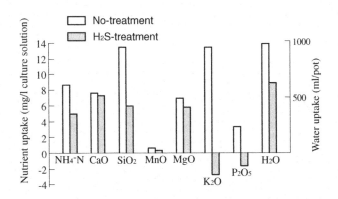

Figure 9.9
Inhibition of nutrient uptake by rice due to hydrogen sulfide
(Source: Mitsui et al., 1951)

upper Myanmar (Tanaka and Yoshida, 1970). Before 1980 there had been only a few reports of sulfur deficiency for rice in India and Indonesia (Ismunadji *et al.* 1975). However recently, sulfur deficiency has been reported more frequently, for example, in Pakistan (Rashid *et al.*, 1992), Bangladesh (Bhuiyan and Islam, 1986), China (Liu, 1986), and the Philippines (Mamaril and Gonzales, 1988). The widespread use of high-analysis fertilizers (such as urea and triple superphosphate) in rice cultivation may have caused this, because they do not contain sulfate as an accessory ingredient.

9.7 Micronutrients

Of the important micronutrient deficiencies and toxicities reported for paddy soils, the toxicities of iron and manganese and deficiencies of manganese, sulfur, and silica have already been addressed. Here, iodine toxicity and zinc and copper deficiencies are discussed.

9.7.1 Iodine toxicity

Iodine toxicity is attributed to a physiological disorder called *akagare* in Japan. *Akagare* is grouped into three types: *akagare* type I is related to a deficiency of K in permanently wet paddy soil and peat soil; *akagare* type II is due to Zn deficiency; and *akagare* type III is caused by iodine toxicity.

Akagare literally means, 'wilting with red coloration'. Type III is also called *kaiden-akagare*, meaning *akagare* associated with newly reclaimed paddy lands because it normally occurs within 2–3 years after upland soils, particularly ando soils (Udands), are reclaimed for rice cultivation.

When soil is kept under upland conditions, iodine is bound strongly by organic matter in the soil, and not readily absorbed by plants. However, once the soil is submerged, iodine is reduced and released from organic matter as I_2 or I^-, and then either absorbed by rice or leached out of the solum. This is the reason why iodine toxicity occurs only in soil newly reclaimed from upland soils. Tensho and Yeh (1970 a,b) clarified the mechanism of the iodine toxicity in *akagare* type III. Figure 9.10 shows the results of one of their experiments showing the dissolution of iodine upon submergence, using [131]I. The added ^{131}I is sorbed by soil organic matter and turns to an insoluble form while the soil is under upland conditions. However, after submergence the iodine is rapidly dissolved from the soil, and this process is accelerated by the addition of readily decomposable organic matter.

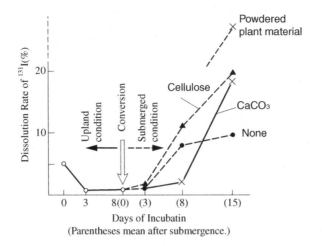

Figure 9.10
Dissolution of iodine upon submergence with addition of various materials
(Source: Tensho and Yeh, 1970a, b)

9.7.2 Zinc deficiency

Japan used to have Zn deficiency, known as *akagare* type II, in rather limited areas. In tropical Asian countries, incidences of Zn deficiency have become a more serious problem since the 1970s. This is related to the fact that more and more land previously unused for rice cultivation in dry climatic regions has been converted to paddy fields as irrigation systems developed. It is estimated that about 2 million ha of paddy soils are deficient in Zn (Randhawa *et al.*, 1978).

The soils that show Zn deficiency are without exception calcareous soils or saline and sodic soils with alkaline pH. Greenland (1997) sets pH 6.8 as a threshold. The solubility of Zn is low in an alkaline soil reaction as shown in Figure 9.11, where the soil was treated with either $Ca(OH)_2$ or Na_2CO_3. In this example, the solubility of Zn may have been governed by either of the following reactions:

$$Zn(OH)_2 = Zn^{2+} + 2OH^- \qquad pK_{sp} = 17.5$$
$$ZnCO_3 = Zn^{2+} + CO_3^{2-} \qquad pK_{sp} = 6.5$$

In a strongly reduced medium where free sulfide occurs, Zn would be precipitated as zinc sulfide:

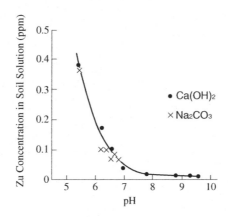

Figure 9.11
Relationship between pH and Zn concentration in the supernatant solution of
a soil treated either with Zn(OH)$_2$ or Na$_2$CO$_3$ (Source: Tanaka et al., 1969)

$$ZnS = Zn^{2+} + S^{2-} \qquad\qquad pK_{sp} = 22.8$$

Yoshida (1970) states that a high concentration of bicarbonate in reduced soil inhibits the translocation of zinc from roots to the top, so aggravating Zn deficiency.

According to the work of the International Rice Research Institute (IRRI, 1981), as well as soils with a high pH, those with a high content of organic matter tend to cause Zn deficiencies in rice. Greenland (1997) sets a threshold value of 3% for organic matter content but this does not appear to hold for many temperate soils. There has been no case of Zn deficiency reported in Japan on paddy soils with high organic matter content, including ando soils that often contain more than 10%. In fact, Japanese paddy soils commonly have an organic matter content of 3% or higher.

The critical Zn content of the plant (dry weight basis) is about 10 mg kg^{-1}. Zinc deficiency can be overcome by dipping seedlings in a 2% suspension of zinc oxide (ZnO). Such a simple treatment can sometimes increase the rice yield by as much as 5 t ha^{-1}.

9.7.3 Copper deficiency

A few papers have reported a substantial increase in rice yields as a result of the soil application and spray application of copper in India (Raheja *et al.*, 1959; Mehrotra and Saxena, 1967) and southern China (Zheng and Huang, 1986; Xu and Dong, 1989).

Copper deficiency is sometimes suspected to be responsible for the poor performance of rice in peat soils. Widjaja-Adhi (1988) reported a case of copper deficiency in rice grown in peat in Indonesia. Using examples from the peat lands in Kalimantan and Sumatra, Driessen (1978) suggested that the sterility of rice in deep peat in the tropics could be attributed to the uncoupling of oxidative phosphorylation in a copper-deficient environment. The copper deficiency in peat soils is ascribed to chelate or complex formation with organic compounds, particularly the polyphenolic compounds that are abundant in peat. Some of the experiments dealing with Cu-related problems in tropical peat are descsribed in Chapter 11.

9.8 *Akiochi* and other fertility problems

Akiochi in Japanese means 'autumnal decline in plant vigor'. In Japan, rice plants show vigorous growth until mid-summer after high doses of fertilizers, but towards the end of summer this growth declines and results in a very unsatisfactory yield. The type of soil that shows the *akiochi* phenomenon is a whitish-colored sandy soil. Its most prominent feature is the presence of a bleached horizon that resembles the albic horizon or E horizon of podzols, although not all of the *akiochi*-affected soils have this feature.

The cause of *akiochi* is linked most directly with the low, free iron oxide content of the soil. High summer temperatures accelerate the decomposition of organic matter, causing a strong reduction in the plow layer. Among the products of anaerobic decomposition, butyric acid and hydrogen sulfide are particularly toxic to rice plants. If the amount of free iron oxide is kept sufficiently high, the *Eh* is maintained at a relatively high level and anaerobic fermentation, which produces butyric acid, and sulfate reduction are suppressed. Therefore, the remedy for *akiochi* soil is, first of all, to incorporate iron-rich red-colored soil material into the plow layer (soil dressing) as already explained in relation to iron.

However, usually, this is not sufficient to improve *akiochi*-affected soil. The low content of free iron oxides indicates the general paucity of nutrients and the poor quality of clay minerals. Kawaguchi and Kyuma (1977) established that the amounts of various nutrients are highly correlated, so soil that is very low in one nutrient tends to be also low in others. Therefore, a low content of iron oxide means, at the same time, low contents available of K, Ca, Mg, Mn and Si and even micronutrients. The phrase 'degraded paddy soils' is, thus, an appropriate term to describe *akiochi*-affected soil. Rice grown on degraded paddy soils frequently displays *helminthosporium* spots, and application of K, Mn, and soluble

SiO_2 are very often effective in improving growth and preventing *akiochi*. The effect of soil dressing is not only to supply iron, but also to supply other nutrients and better quality clay minerals.

References

Aoki, M. 1941. Studies on the behavior of soil phosphoric acid under paddy field condition (I). *J. Sci. Soil & Manure, Japan*, 15: 152–202. (In Japanese).

Bhuiyan, N.I. and Islam, M.M. 1986. Sulfur deficiency problems of wetland rice soils in Bangladesh agriculture. In Portch, S. and Hussain, S.G. (Eds.) *Sulfur in Agricultural Soils*, pp. 85–116. Bangladesh Agric. Res. Council, Dacca, and The Sulfur Inst., Washington, D.C. (Cited from Greenland, 1997).

Cholitkul, W. and Tyner, E.H. 1971. Inorganic phosphorus fractions and their relation to some lowland rice soils of Thailand. *Proc. Int. Symp. Soil Fert. Eval., New Delhi*, 1: 7–20.

Driessen, P.M. 1978. Peat soils. In IRRI (Ed.) *Soils and Rice*, pp. 763–779. IRRI, Los Baños, Philippines.

Furukawa, H. 1978. Transformation of phosphorus and its mobility under submerged condition. In Kawaguchi, K. (Ed.) *Paddy Soil Science*, pp.264–274. Kodansha, Tokyo. (In Japanese).

Furukawa, H. and Kawaguchi, K. 1969. Contribution of organic phosphorus to the increase of easily soluble P in waterlogged soil, especially related to phytic phosphorus (inositol hexa-phosphate) with special reference to phytates. *J. Sci. Soil and Manure, Japan*, 40: 141–148. (In Japanese).

Greenland, D.J. 1997. *The Sustainability of Rice Farming*. CAB Int'l, U.K. and IRRI, Los Baños, Philippines.

Hiroshima Agricultural Experiment Station 1968–1970. Annual Reports. (In Japanese).

Iimura, K. 1982. Chapter 5. Chemistry of Paddy Soils (2). In Yamane, I. (Ed.) *Paddy Soil Science*, pp. 181–232. No-Bun-Kyo, Tokyo. (In Japanese).

Imaizumi, Y. and Yoshida, S. 1958. Silica supplying potential of paddy soils. *Bull. Nat. Inst. Agric. Sci.*, B8: 261–304. (In Japanese).

Inada, K. 1966. Studies on bronzing disease of rice plant in Ceylon. *Proc. Crop Sci. Soc. Jap.*, 33: 309–323.

International Rice Research Institute (IRRI) 1981. *Research Highlights*. IRRI, Los Baños, Philippines. (Cited from Greenland, 1997).

Ismunadji, M., Zulkarnaini, I. and Miyake, M. 1975. *Sulfur Deficiency in Lowland Rice in Java*. Contr. Cent. Res. Inst. Agric., No.14, Bogor.

Iwata, T. 1928. Results of lysimeter studies. *J. Central Agric. Expt. Sta.*, 49: 1–40. (In Japanese).

Kato, N. 1998. Evaluation of silicon availability in paddy soils by an extraction using a phosphate buffer solution. *Summaries of the 16th World Congress of Soil Science, Montpellier, France*, p.266.

Kato, N. and Owa, N. 1990. Dissolution mechanism of silicate slag fertilizers in paddy soil. *Trans. 14th Int'l. Congr. Soil Sci.*, IV: 609–610.

Kato, N., Sumida, H. and Yasuda, M. 2001. Vertical distribution of available silica in paddy soils. *Tohoku Agric. Res.*, 54: 61–62. (In Japanese).

Kawaguchi, K. and Kyuma, K. 1969. *Lowland Rice Soils in Thailand.* Center for Southeast Asian Studies, Kyoto University.

Kawaguchi, K. and Kyuma, K. 1977. *Paddy Soils in Tropical Asia, Their Material Nature and Fertility.* Univ. Press of Hawaii, Honolulu.

Kawaguchi, K. and Matsuo, Y. 1955. Studies on the formation of paddy soils (Part 1). *J. Sci. Soil & Manure, Japan*, 25: 232–236. (In Japanese).

Kawasaki, I. 1953. *Natural Supply Potentials of the Three Major Elements in the Main Arable Lands of Japan.* Nippon Nogyou Kenkyusho, Tokyo. (In Japanese).

Kitada, K., Kamekawa, K. and Akiyama, Y. 1992a. The dissolution of silica in soil in rotational paddy fields by the surface water dissolution method. *Jpn. J. Soil Sci. Plant Nutr.*, 63: 31–38. (In Japanese).

Kitada, K., Kamekawa, K., Akiyama, Y., Shimoda, H. and Yamagata, M. 1992b. Estimation of nutrient movement in rotational paddy soil based on the ripening of upland soil. *Jpn. J. Soil Sci. Plant Nutr.*, 63: 349–351. (In Japanese).

Kogano, K. 1984. Variations in effects of applied phosphate in paddy fields in northern Japan. In Jpn. Soc. Soil Sci. Plant Nutr. (Ed.) *Paddy Soils and Phosphorus*, pp 59–86, Hakuyusya, Tokyo. (In Japanese).

Komoto, Y. 1971. Growth and yield of rice plants in low phosphorus soils. *Japan Agric. Res. Quarterly.(JARQ)*, 6: 63–67.

Kubota, S. 1961. Characteristics of polder land soils and their changes after poldering. *Spec. Rep. Okayama Agric. Expt. Sta.*, No.59. (In Japanese).

Lindsay, W.L. 1979. *Chemical Equilibria in Soils.* John Wiley & Sons, N.Y.

Lindsay, W.L. and Moreno, E.C. 1960. Phosphate phase equilibria in soils. *Soil Sci. Soc. Am. Proc.*, 24: 177–182.

Liu Zhongqun 1986. Preliminary study of soil sulfur and sulfur fertilization efficiency in China. In Portch, S. and Hussain, S.G. (Eds.) *Sulfur in Agricultural Soils*, pp. 371–387. Bangladesh Agric. Res. Council, Dacca, and The Sulfur Inst., Washington, D.C. (Cited from Greenland, 1997).

Mamaril, C.P. and Gonzales, P.B. 1988. Response of lowland rice to sulfur in the Philippines. In *Proc. Int'l Symposium on Sulfur for Korean Agriculture*, pp. 72–76. Korean Soc. Soil Sci. and The Sulfur Inst., Washington, D.C. (Cited from Greenland, 1997).

Mehrotra, O.N. and Saxena, H.K. 1967. Response of important crops to trace elements in Uttar Pradesh. *Indian J. Agron.* 12: 186–192. (Cited from Randhawa *et al.* 1978).

Mitsui, S., Aso, S. and Kumazawa, K. 1951. Dynamic studies on the nutrients uptake by crop plants (Part 1). The effect of butyric acid and respiration inhibitors such as H_2S, NaCN and NaN_3 on the nutrients uptake by rice plant. *J. Sci. Soil & Manure, Japan*, 22: 46–52. (In Japanese).

Mitsui, S., Kumazawa, K. and Ishihara, T. 1953. Dynamic studies on nutrient uptake by crops (Part 7): Effect of the inhibitors to respiratory enzymes, such as hydrogen sulfide, sodium cyanide, sodium azide, etc. and butyric acid on the nutrient uptake by rice roots. *J. Sci. Soil & Manure, Japan*, 24: 45–50. (In Japanese).

Nanzyo, M. 1990. Effect of organic reducing agents on the P sorption products formed on non-crystalline hydrous Fe oxide. *Soil Sci. Plant Nutr.*, 36: 511–514.

Nanzyo, M. 1996. Progress and prospect of the research on paddy soil management under various rice growing systems: 1. Progress in nutrient behavior and management research on paddy soil (2) Phosphorus. *Jpn. J. Soil Sci. Plant Nutr.*, 67: 317–321. (In Japanese).

Nanzyo, M., Takahashi, T. and Shoji, S. 1996. Rapid method to determine available phorphorus content of paddy soils under reducing conditions using ascorbic acid. *Jpn. J. Soil Sci. Plant Nutr.*, 67: 73–77. (In Japanese).

OjimaT. and Kawaguchi, K. 1959. Reaction between difficultly soluble phosphate and hydrogen sulfide; a mechanism of availability and leaching loss of phosphates in paddy soils. *Rep. Food Res. Inst. Kyoto Univ.*, 22: 59–67. (In Japanese).

Osugi, S. and Kawaguchi, K. 1938. On the reduction of sulfates in paddy field soil. *J. Sci. Soil & Manure, Japan*, 12: 453–462. (In Japanese).

Patrick, W.H. 1964. Extractable iron and phosphorus in a submerged soil at controlled redox potentials. *Trans. 8th Int'l. Congr. Soil Sci.*, 4: 605–610.

Patrick, W.H. and Reddy, C.N. 1978. Chemical changes in rice soils. In IRRI (Ed.) *Soils and Rice*, pp. 361–379. IRRI, Los Baños, Philippines.

Raheja, P.C., Yawalkar, K.S. and Singh, R. 1959. Crop response to micronutrients under Indian condition. *Indian J. Agron.* 3: 254–263. (Cited from Randhawa *et al.* 1978).

Randhawa, N.S., Sinha, M.K. and Takkar, P.N. 1978. Micronutrients. In IRRI
(Ed.) *Soils and Rice*, pp. 581–603. IRRI, Los Baños, Philippines.

Rashid,M., Bajwa, M.I., Hussain, R., Naeem-ud-Din, M. and Rehman, F. 1992.
Rice response to sulfur in Pakistan. *Sulfur in Agric.*, 16: 3–5. (Cited
from Greenland, 1997).

Seino, K., Itoh, J., Kosuge, N., Itoh, H. and Yamazaki, T. 1970. Effect of the
mode of nutrient supply and seedling quality on rice productivity in
heavy textured impervious paddy soils. *Bull. Hokuriku Agric. Expt.
Sta.*, 11: 25–82. (In Japanese).

Shiga, H. 1973. Phosphorus fertility and the effect of phosphate fertilization
in paddy soils in the cold region, Part 1. Measurement of phosphorus
supplying potential of paddy soils. *Res.Bull. Hokkaido Nat. Agric.
Expt. Sta.*, 105: 31–49. (In Japanese).

Shiga, H. 1976. Effect of phosphorus fertility of soils and phosphate
application on lowland rice. *Japan Agric. Res. Quarterly. (JARQ)*, 10:
12–16.

Shiga, H. and Suzuki, S. 1963. Behavior of hydrogen sulfide in submerged
soils, Part 6. Iron compounds that react with hydrogen sulfide evolved
in submerged soils. *Bull. Chugoku Nat. Agric. Expt. Sta.*, A9: 197–220.
(In Japanese).

Shiga, H. and Yamaguchi, N. 1977. Studies on the effect of soil phosphorus
and phosphate fertilization on rice cultivation in cool climatic zone.
Part 4. Phosphorus fertility of ordinary paddy soils in Hokkaido. *Res.
Bull. Hokkaido Nat. Agric. Expt. Sta.*, 117: 17–30. (In Japanese).

Shigezumi, M., Kitta, Y. , Kubo, S. and Mizuochi, T. 2002. The evaluation of
the available silica in paddy soil by the phosphate buffer extraction
method. *Jpn. J. Soil Sci. Plant Nutr.*, 73: 383–390. (In Japanese).

Shoji, S. and Mae, T. 1984. III. Dynamics of mineral nutrients and water. In
Ecophysiology of Crops, pp. 97–171. Bun-eido, Tokyo. (In Japanese).

Shoji, S., Yoshida, A., Higuchi, F., Saito, S., Takizawa, A. and Kakizaki,T.
1972. Studies on soils and fertilizers seeking for a breakthrough to the
stagnation of rice yield in the Shonai district: A trial for attaining a
high rice yield based on the improvement of soil drainage. *Rep.
Yamagata Agric. Expt. Sta.*, 6: 42–111. (In Japanese).

Sumida, H. 1991. Characteristics of silica dissolution and adsorption in
paddy soils: Application to soil test for available silica. *Jpn. J. Soil
Sci. Plant Nutr.*, 62: 378–385. (In Japanese).

Sumida, H. 1992. Silica supplying power of paddy soils and silica uptake
characteristics of rice in the cool climate zone. *Bull. Tohoku Nat.
Agric. Expt. Sta.*, 85: 1–46. (In Japanese).

Sumida, H. 1996. Progress and prospect of the research on paddy soil management under various rice growing systems: 1. Progress in nutrient behavior and management research on paddy soil (3) Silicic acid. *Jpn. J. Soil Sci. Plant Nutr.*, 67: 435–439. (In Japanese).

Sumida, H. and Kato, N. 2001. Rice productivity of rotational paddy field converted from upland field cropped with soybean for 18 years. *Tohoku Agric. Res.*, 54: 59–60. (In Japanese).

Sumida, H., Ohyama, N., Nozoe, T. and Sato, T. 1990. Growth, yield and nutrient uptake characteristics of rice and soil nutrient dynamics in the nutrient depletion treatments. *Bull. Tohoku Nat. Agric. Expt. Sta.*, 82: 19–45. (In Japanese).

Suzuki, A. 1997. *Fertilization of Rice in Japan.* Japan FAO Association, Tokyo.

Tadano, T. 1975. Devices of rice roots to tolerate high iron concentration in growth media. *Japan Agric. Res. Quarterly. (JARQ)*, 9: 34~39.

Takagi, S. 1958. Growth of rice on muck soils in the pot and inhibitory factors of its growth. *Rep. Inst. Agric. Res. Tohoku Univ.*, 10: 13–28. (In Japanese).

Takahashi, K. 1981. Studies on fertilizer effect of silicate slag to rice and available silica in soils. *Bull. Shikoku Agric. Expt. Sta.*, 38: 75–114. (In Japanese).

Takahashi, K. and Nonaka, K. 1986. Method of measurement of available silicate in paddy field. *Jpn. J. Soil Sci. Plant Nutr.*, 57: 515–517. (In Japanese).

Takai, Y. Koyama, T. and Kamura, T. 1957 a. Microbial metabolism of paddy soils, Part III. Effect of iron and organic matter on the reduction process (1). *J. Agric. Chem. Soc. Japan*, 31: 211–215. (In Japanese).

Takai, Y. Koyama, T. and Kamura, T. 1957 b. Microbial metabolism of paddy soils, Part IV. Effect of iron and organic matter on the reduction process (2). *J. Agric. Chem. Soc. Japan*, 31: 215–220. (In Japanese).

Takamura, Y. Tabuchi, T. and Kubota, H. 1977 a. Behavior and balance of applied nitrogen and phosphorus under rice field conditions. In *Proc. Int'l Seminar on Soil Environment and Fertility Management in Intensive Agriculture*, pp. 342–349.

Takamura, Y., Tabuchi, T. Harigae, Y., Otsuki, H., Suzuki, S. and Kubota, H. 1977b. Studies on balance sheets and losses of nitrogen and phosphorus in the actual paddy field in the Shintone river basin. *J. Sci.Soil & Manure, Japan*, 48: 431–436. (In Japanese).

Takamura, Y., Tabuchi, T., Suzuki, S., Harigae,Y., Ueno, T. and Kubota, H. 1976. The fates and balance sheets of fertilizer nitrogen and phosphorus applied to a rice paddy field in the Kasumigaura basin. *J. Sci.Soil & Manure, Japan,* 47: 398–405. (In Japanese).

Tanaka, A. and Yoshida, S. 1970. *Nutritional Disorders of the Rice Plant in Asia.* IRRI, Tech. Bull. No. 10, pp. 1–51.

Tanaka, A., Shimono, K. and Ishizuka, Y. 1969. Zinc deficiency as the cause of the *'akagare'* in the rice plant. *J. Sci. Soil & Manure, Japan,* 40: 415–419. (In Japanese).

Tanaka, A., Watanabe, N. and Ishizuka, Y. 1969. A critical study on the phosphate concentration in the soil solution of submerged soils. *J. Sci. Soil & Manure, Japan,* 40: 406–414. (In Japanese).

Tensho, K. and Yeh, K.L. 1970 a. Radio-iodine uptake by plant from soil with special reference to lowland rice, *Soil Sci. Plant Nutr.,* 16: 30–37.

Tensho, K. and Yeh, K.L. 1970 b. Study on iodine and bromine in soil-plant system in relation to the 'reclamation-*akagare*' disease of lowland rice by means of radioisotope-techniques, *Radioisotopes,* 19 (12): 26–31.

Uwasawa, T., Kano, J. and Uchida, Y. 1980. A proposal on quantitative evaluation of soil phosphorus availability. *J. Sci. Soil & Manure, Japan,* 51: 79–84. (In Japanese).

Van Breemen, N. and Moormann, F.R. 1978. Iron-toxic soils. In IRRI (Ed.) *Soils and Rice,* pp. 781–800. IRRI, Los Baños, Philippines.

Widjaja-Adhi, I.P.G. 1988. Physical and chemical characteristics of peat soils of Indonesia. *Indon. Agric.Res. Develop. J.,* 10: 59–64. (Cited from Greenland, 1997).

Xu, J.X and Dong, W.R. 1989. Copper status of permanently and long-term waterlogged paddy soils and response of rice to copper fertilizer. *Acta Pedologica Sinica,* 26:149–158. (Cited from Greenland, 1997).

Yamada, N. 1959. Some aspects of the physiology of Bronzing. *Int. Rice Commission Newsletter,* 8: 11–16.

Yoshida, S. 1961. Natural supply of nutrients by soil and river water. In Konishi, C. and Takahashi, J. (Eds) *Dojo-Hiryo Koza* 1. Asakura-shoten, Tokyo. (In Japanese).

Yoshida, S., McLean, G.W., Shafi, M. and Mueller, K.E. 1970. Effects of different methods of zinc application on growth and yields of rice in a calcareous soil, *West Pakistan. Soil Sci. Plant Nutr.,* 16: 147–149.

Zheng, H.K. and Huang, Z.Q. 1986. Effects of copper on paddy rice and the application technique. *Fujian Agric. Sci. Tech.,* No. 4: 21–22. (Cited from Greenland, 1997)

Chapter 10: Fertility Evaluation and Rating of Paddy Soils in Tropical Asia

10.1 Introduction

Kawaguchi and Kyuma (1977) made an extensive study of paddy soils in tropical Asia to elucidate their material nature and fertility characteristics and ultimately to help increase rice production in the region. More than 500 samples of paddy soils taken from the region were studied in the field and in the laboratory. Although the number is still too small to encompass the entire region in detail, the study was sufficiently broad to allow some generalizations to be drawn. This data assisted the establishment of a scheme for evaluating soil fertility and rating tropical Asian paddy soils.

The greater part of the data used for fertility evaluation in this chapter is introduced in Chapter 2, when describing soil material. Here, the total chemical composition of the soil samples (Kawaguchi and Kyuma, 1977) are also used.

10.2 Various methods used for soil fertility evaluation

10.2.1 Correlation analysis

When many soils are studied in the field and the laboratory, the amount of data obtained is so large that it is almost impossible to make useful conclusions about soil fertility or soil genesis by manual manipulation of the data. It becomes imperative to use a computer to process the data. There are many kinds of multivariate statistical methods for handling such a large data matrix, and the basic prerequisite for these methods is correlations between variables that are obtained for the object of the study—that is, soil. Therefore, the first step in a study of this kind is to examine the correlation matrix of the data used for the analyses.

Table 10.1 presents a matrix of the correlation coefficients for all pairs of twenty-nine variables for the 410 tropical Asian paddy soils studied by Kawaguchi and Kyuma (1977). The variables may be combined into six groups: that is, those related to base status; mechanical composition; clay mineralogical composition; organic matter status; P status; and total chemical composition. To facilitate distinction of the different degrees of correlation, the table has been transformed into a figure by dividing the range of correlation coefficients into five grades, each designated by a specific pattern, as in Figure 10.1. A careful examination of the correlation matrix has revealed several important points.

- Base status characters are highly correlated, not only among themselves, but also with textural composition, clay mineralogy, and part of the total chemical composition. Silt, 1.0 nm minerals, total titanium and total potassium contents are rather exceptional, showing only low to insignificant correlations with base status characters. Of these mutually correlated characters, the sand content of the soil, the 0.7 nm mineral content of the clay fraction, and the total silica content of the soil are negatively correlated with most other characters, although their mutual correlations are positive.
- Characters representing organic matter are not highly correlated with any of the character groups, although their mutual correlations are high.
- This feature also applies to those characters related to available P status.

Such correlations between soil data are basically applicable to any soil groups, although a slight impediment may be expected with intensively managed cultivated soils due to human interference, such as liming and fertilizer application.

10.2.2 Numerical taxonomy

In an attempt to summarize the results of studies on tropical Asian paddy soils, numerical taxonomy was first applied to the data, because the concepts and statistical methods that underlie numerical taxonomy are easy to comprehend.

Numerical taxonomy may be defined as 'the numerical evaluation of the affinity or similarity between taxonomic units and the ordering of these units into taxa on the basis of their affinities' (Sokal and Sneath, 1963). It was originally proposed for the general or natural classification of objects such as plants, insects, and microbes. However, in this study, the method is used to create groups on the basis of the similarities between a limited number of characters that are relevant to soil fertility.

The procedure of numerical taxonomy consists of the following steps:

Step 1—standardization of the data to make them dimensionless;

Step 2—computation of between-sample similarity coefficients;

Table 10.1 Correlation coefficient matrix between all pairs of 29 variables, with decimal points omited (Source: Kawaguchi and Kyuma, 1977)

	pH	Ex-Ca	Ex-Mg	Ex-(Ca+Mg)	Ex-Na	Ex-K	CEC	Avail. SiO_2	Sand	Silt	Clay	0.7 nm Min.	1.0 nm Min.	1.4 nm Min.	TC	TN	NH_4-N	TP	Bray-P	HCl-P	TSIO	TFEO	TALO	TCAO	TMGO	TMNO	TTIO	TKAO
pH																												
Ex-Ca	622																											
Ex-Mg	289	534																										
Ex-(Ca + Mg)	569	943	786																									
Ex-Na	298	275	579	430																								
Ex-K	324	448	633	577	609																							
CEC	399	855	781	934	371	595																						
Avail. SiO_2	625	701	527	721	293	514	718																					
Sand	-016	-463	-524	-545	-243	-456	-662	-351																				
Silt	-041	-027	016	-014	-032	015	014	-074	-564																			
Clay	045	577	623	668	313	542	791	470	-854	052																		
0.7 nm Min.	-537	-564	-474	-600	-236	-378	556	-412	283	-209	-211																	
1.0 nm Min.	093	-125	-164	-156	-023	-221	-066	-302	-066	280	-096	-230																
1.4 nm Min.	462	639	585	699	253	330	688	527	-282	330	585	-759	-325															
TC	-254	002	152	062	044	139	233	070	-316	110	313	-008	-159	040														
TN	-264	-057	085	-008	022	122	154	026	-297	166	255	-018	-105	-012	958													
NH_4-N	-117	009	084	040	021	115	143	222	-190	108	162	008	-305	100	514	586												
TP	245	155	203	148	320	244	415	080	-198	135	154	-201	-023	080	292	333	323											
Bray-P	268	079	026	068	145	039	457	-057	078	244	-059	-144	166	038	010	009	-047	806										
HCl-P	379	154	057	136	257	356	047	159	094	-001	-113	-290	253	114	-007	002	-059	333	436									
TSIO	-322	-512	-502	-573	-218	-456	-631	-603	655	-223	-651	-326	002	-209	-212	-209	-243	-509	-020	-119								
TFEO	389	557	504	607	202	384	638	685	-512	100	555	-304	045	373	045	013	114	481	-008	066	-841							
TALO	066	436	429	162	404	544	712	000	-716	244	457	-198	183	330	330	321	192	432	010	210	-919	657						
TCAO	553	418	157	368	109	174	278	434	-010	018	000	-449	402	183	-050	065	089	243	075	272	-406	289	159					
TMGO	494	434	415	328	174	482	328	434	-324	226	250	-501	110	377	038	065	099	342	102	144	-627	502	411	600				
TMNO	404	466	431	345	331	511	482	434	-262	039	292	-249	-239	376	-069	-107	014	393	049	099	-525	686	351	300	350			
TTIO	102	195	241	238	095	274	231	-166	-166	-066	242	-065	-007	158	004	049	-007	201	-024	007	-350	499	217	-006	150	385		
TKAO	136	-014	-017	052	158	-066	018	-128	-133	227	018	-234	666	-131	004	049	-153	036	183	285	-224	-042	192	-015	260	-186	-113	
TPHO	173	193	144	198	269	239	380	109	-118	054	109	-216	-082	073	344	376	387	734	249	357	471	392	409	278	275	333	224	043

Notes: Abbreviations for variable names are as in the text; TSIO to TPHO are for total elemental oxides in the order of SiO_2, Fe_2O_3, AL_2O_3, CaO, MgO, MnO_2, TiO_2, K_2O and P_2O_5. (Source: Kawaguchi and Kyuma, 1977)

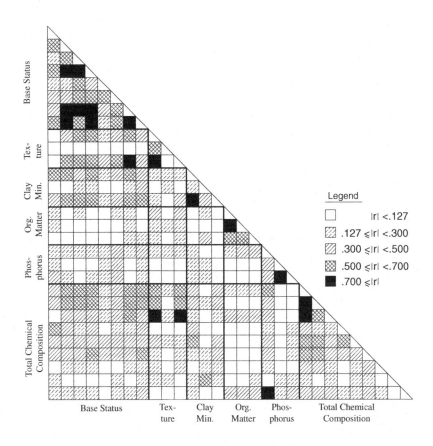

Figure 10.1
Patternized expression of the correlation matrix in Table 10.1
(Source: Kawaguchi and Kyuma, 1977)

Step 3—sorting or clustering;

Step 4—formulating a dendrogram.

Usually, a between-sample correlation coefficient (r_{jk}) and a taxonomic distance or Euclidean distance (d_{jk}) are used as similarity coefficients to represent similarities in the pattern and magnitude, respectively.

$$r_{jk} = \frac{\sum_{i=1}^{n}(X_{ij} - \bar{X}_j)(X_{ik} - \bar{X}_k)}{\sqrt{\sum_{i=1}^{n}(X_{ij} - \bar{X}_j)^2 \sum_{i=1}^{n}(X_{ik} - \bar{X}_k)^2}}$$

$$d_{jk} = \left[\frac{\sum_{i=1}^{n} (X_{ij} - X_{ik})^2}{n} \right]^{\frac{1}{2}}$$

As a representative sorting or clustering method, the weighted pair-group method was used. This method allows the two most mutually similar operational taxonomic units (a soil or a soil group) to join in one clustering cycle.

An example of a dendrogram is shown in Figure 10.2. The results of numerical taxonomy were satisfactory, in that soils apparently similar in the field observations were combined in a cluster with a high similarity coefficient. However, useful as it was for summarizing the soil data, numerical taxonomy could not show the fertility relationship among the established clusters, that is, it could not show which soils were more or less fertile relative to others. Thus, it was necessary to search for a method that could reveal fertility relationships and could help rate the sample soils. However, numerical taxonomy can be effectively used to classify climate and soil material (Kyuma, 1972; Kyuma *et al.*, 1974; Kawaguchi and Kyuma, 1977).

In the course of manipulating the data to refine the numerical taxonomy results, it was found that principal component analysis–used to extract mutually independent compound characters–rated the chemical potentiality of the soil samples. Thus, further studies made along this line discovered that the method of factor analysis was promising as a refinement of principal component analysis for making the numerical evaluation of soil fertility for tropical Asian paddy soils.

10.2.3 Principal component analysis and factor analysis

Principal component analysis is used to attain a 'parsimonious summarization of a mass of observation' (Seal, 1964). It extracts the hidden essence of a thing or material that is not directly measurable. Soil fertility is the essence being sought, but it cannot be measured for itself. Measurements are taken of many analytical items of a soil that are related to fertility in different ways and in different degrees. Thus, the model of principal component analysis appears to fit aptly to the study of soil fertility (Kyuma and Kawaguchi, 1973).

Given *n* samples, each of which is defined by *p* characters, the samples can be expressed as *n* points scattered in *p*-dimensional space. The principal component analysis aims to reduce the *p*-axes to orthogonal *m*-axes, where

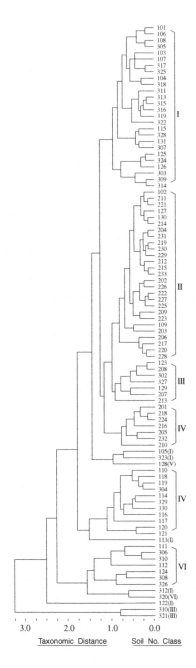

Figure 10.2
A dendrogram showing relationship among ninety-four surface soils with respect to
fertility based on taxonomic distance (Source: Kyuma, 1981)

$m < p$, with a minimal loss of information. Mathematically, this produces a set of new m variables from the original p variables by orthogonal transformation.

To illustrate the principle of the principal component analysis, a schematic diagram of the case with two variables is shown in Figure 10.3. When the two variables, X_1 and X_2, are highly correlated, the axes can be rotated to the position of Y_1 and Y_2, so that the variance along the Y_1 axis becomes maximum and that along the Y_2 axis is minimum. If the latter is sufficiently small, Y_2 can be neglected and Y_1 stands alone as a compound character of X_1 and X_2. In this way, the number of axes is reduced from two to one with a minimum of loss of information.

The new compound variables Y_1 and Y_2 may be expressed linearly as:

$$Y_1 = a_{11}X_1 + a_{12}X_2$$
$$Y_2 = a_{21}X_1 + a_{22}X_2$$

in terms of X_1 and X_2 multiplied by coefficients, a_{11}, a_{12}, and a_{21}, a_{22}, which are called factor loadings.

Factor analysis has aims and procedures similar to those of principal component analysis. However, there are certain differences. It is not necessary for principal component analysis to have any previous assumptions concerning the number and character of the principal components or factors to be extracted, whereas factor analysis must assume the following (Asano, 1971; Kyuma, 1973a):

• the number of common factors to be considered;
• the extent of the contribution of each variable to the common factors.

The fundamental model of factor analysis is expressed as follows:

$$X_i = a_{i1}f_1 + a_{i2}f_2 + \ldots + a_{ik}f_k + \ldots + a_{im}f_m + e_i$$

where f_k ($k = 1, 2, \ldots, m$) is the score of m factors for each sample, $\{a_{ik}\}$ ($i = 1, 2, \ldots, p$; $k = 1, 2, \ldots, m$) is the factor loading for the ith variable and k^{th} common factor, and e_i ($1, 2, \ldots, p$) is the error term or specific character of each variable that is not explained by the m common factors. Given n samples with p variables, factor analysis aims to simultaneously determine the best estimates of a factor loading matrix and an error variance based upon certain assumptions.

Factor loading $\{a_{ik}\}$ may be regarded as a correlation coefficient between the ith variable and the kth common factor. Therefore, the degree of contribution of a variable to a common factor can be judged from the

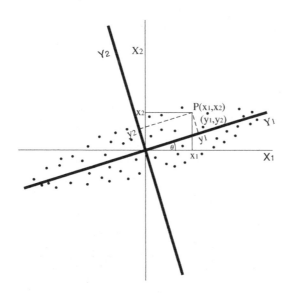

Figure 10.3
Schematic diagram showing the principle of principal component analysia
(Source: Kyuma, 1981)

factor loading. This gives the basis for interpreting the common factors. Another feature of factor analysis is that the estimated factor axes can be rotated freely, so as to make interpretation of the factor easier. This is possible because of what is called the 'indeterminacy' of factor axes. If the factors, after rotation, are interpretable, the computation of factor scores (f_k in the model) can follow. There are several methods available for computing factor scores, but the principle underlying these is the least square estimation.

10.3 Fertility evaluation for tropical Asian paddy soils

10.3.1 Procedures to attain numerical evaluation of soil fertility

In the study evaluating the fertility of 410 tropical Asian paddy soils (Kawaguchi and Kyuma, 1977), the eleven characters or variables listed in Table 10.2 were selected after many preliminary trials. The selection was made considering the following:
- the preference for primary data directly obtained from analysis.
- the omission of an item of complementary data.

The second point means that one of the characters that were mutually highly negatively correlated was omitted, leaving the character that was easier to interpret. For example, clay and sand are highly negatively correlated, and a high sand content in soils has a clearly negative effect on fertility. However, a soil with high clay content does not always have a positive effect. Therefore, sand was used in the analysis.

Since a normal distribution of the variables is implicitly assumed in factor analysis, logarithmic transformation was used to make the variables with positively skewed distribution, such as Bray-P and HCl-P, approach the normal distribution. Based on the preliminary results of the principal component analysis, three factors were extracted that represent more than 80% of the total variance or information of the eleven variables.

Table 10.2 List of characters or variables used for analysis

Character No.	Name	Brief description
X_1	TC (total carbon)	as % of air-dried soil, Tyurin's wet combustion method
X_2	TN (total nitrogen)	as % of air-dried soil, Kjeldahl digestion and steam distillation
X_3	NH_3-N	in mg N $100g^{-1}$ of air-dried soil, after incubation for 2 weeks at 40°C
X_4	TP (total phosphorus)	in mg P_2O_5 100 g^{-1} of air-dried soil, HF-H_2SO_4 or HNO_3-H_2SO_4 digestion
X_5	Bray-P	in mg P_2O_5 100 g^{-1} of air-dried soil, Bray-Kurtz No. 2 method
X_6	HCl-P	in mg P_2O_5 100 g^{-1} of air-dried soil, $0.2N$ HCl extraction at 40°C for 5 hours
X_7	Ex-Ca+Mg	in cmol kg^{-1} of air-dried soil, N NaCl extraction, EDTA titration
X_8	Ex-K	in cmol kg^{-1} of air-dried soil, N NH_4-acetate extraction, flame photometry
X_9	CEC	in cmol kg^{-1} of air-dried soil, buffered neutral N $CaCl_2$ medium
X_{10}	Av-Si (available silica)	in mg 100 g^{-1} of air-dried soil, pH4 acetate buffer extraction at 40°C
X_{11}	Sand	as % of organic matter-free dried soil, sum of coarse and fine sands

(Source: Kawaguchi and Kyuma, 1977)

In the factor analysis that followed the best estimate of the factor loading matrix for the predetermined number of factors (three factors) was obtained after many iterative computations. This factor loading matrix further underwent a rotation of factor axes, using Kaiser's varimax rotation, to provide the terminal factor loading matrix, as shown in Table 10.3.

The first factor is highly correlated only with the characters related to base status and parent material (such as CEC, exchangeable cations, available silica, total P and sand), with the sand being opposite in sign to the rest. Thus, the first factor is called the 'inherent potentiality' (IP), and is determined primarily by the nature and amount of clay and base status.

The second factor is related to TC, TN, and NH_3-N. Moderately high factor loadings on TP and Sand are interpretable in terms of organic P and the textural control of organic matter accumulation, respectively. Therefore, the second factor may be called 'organic matter and N status (OM)' and considered to express the N supplying capacity of soil.

The third factor can clearly be interpreted as 'available P status (AP)'. Factor loading on TP is much less than those on Bray-P and HCl-P. The contribution of other variables to this factor is minor.

It was observed that these three mutually orthogonal, or independent, factors accord with the result of the correlation analysis referred to in the preceding section. Characters related to both the organic matter status and the P status are highly correlated within the respective groups, but rather indifferent to other characters related to base status, textural

Table 10.3 Terminal factor loading matrix for three factors after varimax rotation

	Factor 1	Factor 2	Factor 3
(TC)	0.288	0.913	0.118
(TN)	0.173	0.965	0.151
(NH_3-N)	0.113	0.668	0.046
(TP)	0.478	0.407	0.531
(Bray-P)	0.089	0.104	0.943
(HCl-P)	0.182	0.042	0.881
(Ex-Ca+Mg)	0.936	0.276	0.108
(Ex-K)	0.944	0.053	0.147
(CEC)	0.777	0.246	0.303
(Av-Si)	0.796	0.118	0.218
(Sand)	−0.486	−0.323	0.086

Notes: Character symbols in parentheses denote log-transformed characters.
(Source: Kawaguchi and Kyuma, 1977)

Table 10.4 Factor score coefficient matrix for three factors

	Factor 1	Factor 2	Factor 3
(TC)	−0.151	0.268	−0.078
(TN)	−0.147	0.839	0.010
(NH$_3$-N)	0.045	−0.012	0.008
(TP)	0.051	−0.025	0.084
(Bray-P)	−0.091	−0.101	0.701
(HCl-P)	−0.059	−0.033	0.278
(Ex-Ca+Mg)	0.306	−0.144	0.029
(Ex-K)	0.130	0.018	−0.026
(CEC)	0.757	−0.132	−0.214
(Av-Si)	−0.058	0.073	0.087
(Sand)	0.028	0.012	0.004

Notes: Character symbols in parentheses denote log-transformed characters.
(Source: Kawaguchi and Kyuma, 1977)

composition, and clay mineralogical composition. This leads to interesting and important inferences that the fertility of tropical Asian paddy soils has at least three major components, and that both the OM and AP of these soils are independent of their inherent potentiality (IP).

After interpreting the three factors, their scores were computed for individual soil samples so that a quantitative evaluation of the three components of soil fertility could be made. The coefficient matrix for the score computation is provided in Table 10.4. Since the data were log-transformed before the analysis, the same transformation is needed for score computation. Moreover, the transformed data must further be standardized using the mean and standard deviation vectors given in Table 10.5. The scores are the sum of the products of the coefficient, and the transformed and standardized datum corresponding to the co-efficient[1].

The scores thus calculated for the samples are already standardized with a mean of zero and a variance of unity. Positive score values indicate an above-average status for the overall mean of the 410 sample soils, while negative values indicate a below-average status.

All these calculation procedures for numerical fertility evaluation were developed using the data from the 410 samples of tropical Asian paddy soils taken before 1975. After establishing the method, many more samples were collected from different areas of the same region, such as the Mekong Delta of Vietnam, Sarawak in East Malaysia and the Irrawaddy Delta of Myanmar. Based on the assumption that these latter samples belong to the

Table 10.5 *Means and standard deviations of the eleven log-transformed characters*
for 410 samples

	Mean	Standard deviation
(TC)	0.044	0.297
(TN)	−0.994	0.282
(NH$_3$-N)	0.731	0.425
(TP)	1.775	0.429
(Bray-P)	0.171	0.584
(HCl-P)	0.608	0.712
(Ex-Ca+Mg)	0.993	0.484
(Ex-K)	−0.623	0.449
(CEC)	1.159	0.342
(Av-Si)	1.195	0.515
(Sand)	1.314	0.544

Notes: Character symbols in parentheses denote log-transformed characters.
(Source: Kawaguchi and Kyuma, 1977)

same population as the previous samples, the equations for computing the
scores of IP, OM, and AP were applied to the latter data. Thus, the total
number of soil samples was expanded to 529. The tables in Chapter 2 that
describe paddy soil material in tropical Asia contain these additional
samples (Kyuma, 1985).

The scores of soil fertility components, that is, IP, OM, and AP, may be
utilized to predict rice yields. In recent years, many attempts have been
made to derive a crop-yield prediction equation using multiple regression
analysis, in which soil characters are used as independent variables,
together with such factors as climate and management that are thought to
be relevant to crop yield. Soil characters used in such attempts are often
humus content, clay content, and specific nutrient contents. The three
factor scores evaluated in the above procedure are considered to be most
suitable for the purpose, because:
- the three factors are compound characters derived from many
  individual characters and represent the most important fertility
  components relevant to the yield.;
- these factors are mutually independent, so they best fit the model of
  the multiple regression analysis.

Kyuma (1973 b) showed in a study using survey data from West Malaysia
that these three factors alone accounted for about 60% of the variation of
paddy yield. In that study, IP was shown to contribute the most to paddy
yield of the three factors.

10.3.2 Rating of soil fertility by countries and regions

Although there are some reservations about the sampling procedure adopted in this study, the fertility status of each country can be roughly estimated by calculating the mean of each factor for the samples taken from each country. Table 10.6 shows these values.

IP is highest for the soils of Indonesia and the Philippines, followed by those of India. Indonesia and the Philippines are in a region influenced by active volcanism and the parent material of the soil is continuously rejuvenated by fresh volcanic ejecta. India is located in semi-arid to subhumid climatic regions, so the weathering and leaching of its soils have not been very intensive, especially in the basaltic rock area of the Deccan Plateau that constitutes the catchment of the Godavari and Krishna rivers.

The soils of West Malaysia and Sri Lanka, which are situated in permanently humid to monsoonal climatic regions of the low latitudes, are among the poorest with respect to IP. The soils of Bangladesh, Cambodia and Thailand are mostly on the poorer side of the overall mean.

OM is by far the highest for Malaysian soils, with a mean as high as 1.40. The second highest is the Philippines, with a mean of 0.34. Conversely Indian soils, with the lowest mean of −0.73, are the poorest. The soils of the other countries are more or less similar, with mean scores clustering around

Table 10.6 Means and standard deviations of the three factor scores for the respective countries

Country	No. of samples	IP Mean	IP Std. dev.	OM Mean	OM Std. dev.	AP Mean	AP Std. dev.
Bangladesh	53	−0.438	0.708	0.176	0.704	0.459	0.800
Cambodia	16	−0.231	0.990	−0.155	0.954	−1.277	0.811
India	73	0.449	0.837	−0.780	0.619	0.581	0.906
Indonesia	44	0.618	0.766	−0.014	0.746	0.031	0.734
Malaysia, E.	36	−0.356	0.392	1.547	0.034	−0.643	0.464
Malaysia, W.	41	−0.545	0.719	1.398	0.927	0.026	0.726
Myanmar	50	0.416	0.519	0.102	0.581	−0.798	1.007
Philippines	54	0.618	0.703	0.337	0.673	0.022	1.041
Sri Lanka	33	−0.510	0.853	0.150	1.064	−0.110	0.650
Thailand	80	−0.364	1.195	−0.347	0.939	−0.641	0.769
Vietnam	49	0.050	0.464	0.920	0.832	−0.412	0.557

Notes: Myanmar—mainly from the Irrawaddy Delta; India—from the states along the Ganges and the east coast of Deccan; Indonesia—Java; E. Malaysia—Sarawak; Vietnam—Mekong Delta; for others more or less from throughout the country. (Source: Kyuma, 2001)

the overall mean. It seems that high OM scores are associated with a humid climate and fine soil texture, in addition to a swampy terrain.

AP is high for the soils of India and Bangladesh, while the soils of Cambodia and Thailand are the poorest. The regionality observed in this property is difficult to explain, but may be ascribed to the nature and degree of the weathering of the parent rocks.

Similar calculations were made for the regions that can be defined, more or less discretely, by climate, parent material and geographical area. (see Table 10.7). The highest scores for IP and AP are for the soils of the Godavari-Krishna Delta region, which also have the second lowest mean OM score. The soils of the Northeast Plateau region of Thailand are characterized by very low scores of all three fertility components, reflecting the very siliceous, sandy nature of their parent material.

10.4 Fertility classification and mapping

In order to prepare a fertility classification of the sample soils, the whole range of computed scores was divided into classes, with arbitrary class limits of ±0.25 and ±0.84. The assumption underlying the selection of limits is that, if the distribution of the scores is normal, five classes of almost equal size should be created. The potentiality of each fertility component at different class levels can be designated as follows:

Class No.	Class Limits	Potentiality
1	>0.84	very high
2	0.84 ~ 0.25	high
3	0.25 ~ −0.25	intermediate
4	−0.25 ~ −0.84	low
5	−0.84 >	very low

Although clay mineral composition and total chemical composition were not directly used in the computation of fertility component scores, they are well represented by inherent potentiality, or IP. This is clear from Table 10.8, which shows the mean contents of clay mineral species and selected elemental oxides for each of the five IP classes. The difference in the means among the five classes is statistically highly significant (F values are much higher than the criterion value of $F = 2.4$ at a 5% probability level). Of the variables listed in the table, 1.0 nm mineral and total potash content have a peculiar pattern, with their maximum in the intermediate class.

Table 10.7 Means and standard deviations of the three factor scores for selected regions

Regions	No. of samples	IP Mean	IP Std. dev.	OM Mean	OM Std. dev.	AP Mean	AP Std. dev.
SOUTH ASIA							
India							
Godavari-Krishna Delta	10	1.38	0.29	-0.73	0.47	1.11	0.95
Bangladesh							
Ganges Delta	15	0.33	0.35	0.04	0.82	0.90	0.50
Brahmaputra Delta	13	-0.57	0.48	-0.01	0.64	1.01	0.66
Madhpur-Barind Tracts	9	-0.75	0.60	0.35	0.64	-0.14	0.40
Peripheral Areas	16	-0.87	0.62	0.36	0.66	-0.06	0.77
Sri Lanka							
Wet & Intermed. Zones	14	-1.07	0.64	0.84	1.10	-0.10	0.64
Dry Zone	19	-0.10	0.76	-0.36	0.70	-0.12	0.67
SOUTHEAST ASIA							
Thailand							
Northeast Region	32	-1.18	1.27	-1.14	0.77	-0.94	0.70
Upper Central Plain	14	-0.04	0.76	-0.05	0.49	-0.31	0.69
Bangkok Plain	24	0.53	0.64	0.24	0.66	-0.51	0.86
Peninsular Region	6	-0.40	0.74	0.58	0.48	-0.29	0.39
Cambodia							
Central High Land	8	-0.91	0.76	-0.60	0.94	-1.19	0.69
West Malaysia							
Kedah-Perlis Plain	10	0.25	0.44	1.21	0.56	0.02	0.54
East Coast	10	-1.23	0.20	0.78	0.59	-0.54	0.57
Indonesia							
Central & East Java	28	0.91	0.59	-0.29	0.53	0.10	0.81
Philippines							
Central Plain of Luzon	25	0.91	0.54	-0.10	0.41	-0.36	0.97

Notes: Myanmar—mainly from the Irrawaddy Delta; India—from the states along the Ganges and the east coast of Deccan; Indonesia—Java; E. Malaysia—Sarawak; Vietnam—Mekong Delta; for others more or less from throughout the country. (Source: Kyuma, 2001)

The fertility class number of each sample was plotted in a map at the corresponding sampling site. A set of three maps for IP, OM and AP, respectively, was prepared for each country. Figure 10.4 shows maps of Java, Indonesia which indicate problem areas for each of the three fertility components.

Notes

(1) The equations for the computation of the three fertility component scores are as follows (for X_1–X_{11}, see Table 10.2):

$$IP = -0.151(\log X_1 - 0.044)/0.297 - 0.147(\log X_2 + 0.994)/0.282 +$$

Table 10.8 Mean contents of clay mimeral species and selected elemental oxides for the samples in each inherent potentiality class

IP class	1	2	3	4	5	F-value
No. of samples	88	92	77	70	83	
0.7 nm mineral	27.84	40.27	44.03	52.71	67.83	51.34
1.0 nm mineral	8.13	15.00	17.60	18.00	11.15	7.95
1.4 nm mineral	64.03	44.73	38.38	25.50	19.82	78.09
SiO_2	63.86	66.80	71.61	74.94	84.89	73.46
Fe_2O_3	9.24	7.42	5.75	4.52	2.26	74.89
Al_2O_3	20.31	19.31	16.22	15.60	9.78	41.84
CaO	2.13	1.72	1.69	1.09	0.37	11.25
MgO	1.25	1.20	1.09	0.68	0.31	28.14
TiO_2	1.36	1.25	1.13	1.07	0.87	8.91
K_2O	1.50	2.00	2.28	1.90	1.53	5.90

(Source: Kawaguchi and Kyuma, 1977)

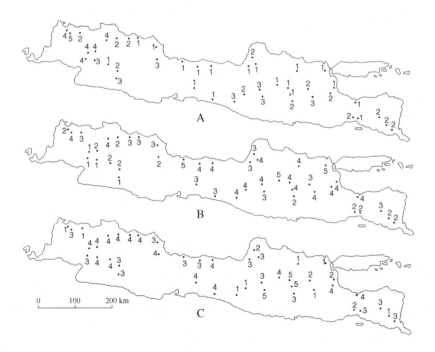

Figure 10.4
Map of Java, Indonesia, showing distribution of paddy soil samples in terms of: A, inherent potentiality class; B, organic matter and nitrogen class; C, available phosphorus class (Source: Kawaguchi and Kyuma, 1977)

$$0.045(\log X_3 - 0.731)/0.425 + 0.051(\log X_4 - 1.175)/0.429 -$$
$$0.091(\log X_5 - 0.171)/0.584 - 0.059(\log X_6 - 0.608)/0.712 +$$
$$0.306(\log X_7 - 0.993)/0.484 + 0.130(\log X_8 + 0.623)/0.449 +$$
$$0.757(\log X_9 - 1.159)/0.342 - 0.058(\log X_{10} - 1.195)/0.515 +$$
$$0.028(\log X_{11} - 1.314)/0.544$$

$$OM = 0.268(\log X_1 - 0.044)/0.297 + 0.839(\log X_2 + 0.994)/0.282$$
$$- 0.012(\log X_3 - 0.731)/0.425 - 0.025(\log X_4 - 1.175)/0.429 -$$
$$0.101(\log X_5 - 0.171)/0.584 - 0.033(\log X_6 - 0.608)/0.712 -$$
$$0.144(\log X_7 - 0.993)/0.484 + 0.018(\log X_8 + 0.623)/0.449 -$$
$$0.132(\log X_9 - 1.159)/0.342 + 0.073(\log X_{10} - 1.195)/0.515 +$$
$$0.012(\log X_{11} - 1.314)/0.544$$

$$AP = -0.078(\log X_1 - 0.044)/0.297 + 0.010(\log X_2 + 0.994)/0.282$$
$$+ 0.008(\log X_3 - 0.731)/0.425 + 0.084(\log X_4 - 1.175)/0.429 +$$
$$0.701(\log X_5 - 0.171)/0.584 + 0.278(\log X_6 - 0.608)/0.712 +$$
$$0.029(\log X_7 - 0.993)/0.484 - 0.026(\log X_8 + 0.623)/0.449 -$$
$$0.214(\log X_9 - 1.159)/0.342 + 0.087(\log X_{10} - 1.195)/0.515 +$$
$$0.004(\log X_{11} - 1.314)/0.544$$

References

Asano, C. 1971. *An Introduction to Internal and External Factor Analyses.* Kyoritsu-Shuppan, Tokyo. (In Japanese).

Kawaguchi, K. and Kyuma, K. 1977. *Paddy Soils in Tropical Asia,Their Material Nature and Fertility.* Univ. Press of Hawaii, Honolulu.

Kyuma, K. 1972. Numerical classification of climate of south and Southeast Asia. *Tonan Ajia Kenkyu (Southeast Asian Studies)*, 9: 502–521.

Kyuma, K. 1973a. A method of fertility evaluation for paddy soils, II. Second approximation: Evaluation of four independent constituents of soil fertility. *Soil Sci. Plant Nutr.*, 19: 11–18.

Kyuma, K. 1973b. A method of fertility evaluation for paddy soils, III. Third approximation: Synthesis of fertility constituents for soil fertility evaluation. *Soil Sci. Plant Nutr.*, 19: 19–27.

Kyuma, K. 1981. Fertility of paddy soils in tropical Asia. *Proc. Symp. on Paddy Soils, Nanjing*, pp. 118–128.

Kyuma, K. 2001. I-2. Lowland soils in Tropical Asia. In Kyuma, K. (Ed.) *Tropical Soil Science*, pp. 157–180. Nagoya Univ. Press, Nagoya. (In Japanese).

Kyuma, K. and Kawaguchi, K. 1973. A method of fertility evaluation for paddy soils, I. First approximation: Chemical potentiality grading. *Soil Sci. Plant Nutr.*, 19: 1–9.

Kyuma, K., Okagawa, N. and Kawaguchi, K. 1974. A numerical approach to the classification of alluvial soil materials. *Trans. 10ᵗʰ Int'l Congr. Soil Sci., Moscow*, VI (2): 543–551.

Seal, H.L. 1964. *Multivariate Statistical Analysis for Biologists*. Methuen & Co., London.

Sokal, R.R. and Sneath, P.H.A. 1963. *Principles of Numerical Taxonomy*. Freeman & Co., San Francisco.

Chapter 11: Problem Paddy Soils

11.1 Acid sulfate soils

Mangroves occur in many coastal areas in tropical Asia but are considered to have originated in Southeast Asia (Yamada, 1986). They consist of many tree species that can tolerate marine and brackish water environments, and they protect shorelines from wave action. They provide spawning and hatching grounds for fish and prawns, and the raw material for making charcoal. Their well-developed root systems create not only a special kind of landscape, but also their strategic location is important for accelerating mud sedimentation and seaward land accretion. In their natural state, mangroves are a very important element of coastal ecosystems in tropical and subtropical regions. However, once they undergo a change that exposes mangrove mud permanently to the air, such as drainage and land reclamation, all of the biota are profoundly affected by the chemical changes of their habitat; that is, by the formation of acid sulfate soils. This change in the condition of the ecosystem is probably one of the most regressive forms of land degradation, both in time and space.

11.1.1 Chemical reactions leading to acid sulfate soil formation

Acid sulfate soils are formed by the oxidation of sulfide-containing sediments, upon their exposure to the atmosphere, in the course of natural land accretion or artificial land reclamation processes. The general conditions that lead to the formation of sulfide-containing sediments are:
- the presence of marine or brackish water as the source of sulfate;
- stagnant water, as typically found in lagoons and bays;
- the supply of decomposable organic matter.

The third condition is also favored by the second, with the stagnant water facilitating the establishment of vegetation, such as mangroves. Consequently, the slow accretion of coastal land by silting may be considered as another condition that favors the formation of sulfide-containing sediments.

The chemical process for the formation of sulfidic compounds in mangrove mud proceeds as follows:

$$Fe(OH)_3(s) + 3H^+ + e = Fe^{2+} + 3H_2O \quad \text{Dissolution of ferrous iron}$$
$$SO_4^{2-} + 8H^+ + 8e = S^{2-} + 4H_2O \quad \text{Sulfate reduction}$$
$$Fe^{2+} + S^{2-} = FeS (s) \quad \text{Precipitation of ferrous sulfide}$$
$$FeS(s) + S(0) = FeS_2(s) \quad \text{Pyrite formation}$$
$$[S(0): \text{elemental sulfur}]$$

It is known that most of the oxidizable sulfide compounds are present in the form of pyrite, FeS_2, because of its relative stability over ferrous sulfide, FeS. In the process of sedimentary pyrite formation, not only sulfate reduction, but also the formation of elemental sulfur must precede. Van Breemen and Pons (1978) consider that a partial oxidation of sulfide to elemental sulfur occurs. Such a predominantly anaerobic process, alternating with a limited aerobic process, can best be met in the zone of tidal fluctuation. However, according to Stumm and Morgan (1970) under acid conditions, elemental sulfur is formed by the reduction of SO_4^{2-} as an intermediate product, and once formed, it may persist as a stable solid. Thus, solid elemental sulfur occurs quite commonly in recent marine sediments.

Upon exposure to the air, pyrite undergoes oxidation as follows:

$$FeS_2(s) + (7/2)O_2 + H_2O = Fe^{2+} + 2SO_4^{2-} + 2H^+ \tag{1}$$

The ferrous iron is further oxidized to ferric iron and this is precipitated if the environmental pH is higher than about 3.

$$Fe^{2+} + (1/4)O_2 + H^+ = Fe^{3+} + (1/2)H_2O \tag{2}$$
$$Fe^{3+} + 3H_2O = Fe(OH)_3(s) + 3H^+ \tag{3}$$

Thus, the overall reaction is:

$$FeS_2(s) + (15/4)O_2 + (7/2)H_2O = Fe(OH)_3(s) + 2SO_4^{2-} + 4H^+ \tag{4}$$

producing four equivalents of acidity from the oxidation of one mole of pyrite.

It is known that the reaction (2) is a slow process if it occurs in a purely chemical fashion. At pH 3, the half time of this reaction is of the order of 1000 days (Stumm and Morgan, 1970), but in the soil, this reaction is mediated by autotrophic iron bacteria, *Thiobacillus ferrooxidans* and *Ferrobacillus ferrooxidans*, and proceeds much faster.

Another important reaction is the oxidation of pyrite by Fe^{3+},

$$FeS_2 + 14Fe^{3+} + 8H_2O = 15Fe^{2+} + 16H^+ + 2SO_4^{2-}$$

which produces more acidity. This reaction runs quite rapidly, and the half time of this reaction is of the order of 20 to 1000 minutes. The oxidation of sulfur in this reaction is mediated by another autotrophic bacteria, *Thiobacillus thiooxidans*.

Stumm and Morgan (1970) provided the schematic drawing of the overall process of pyrite oxidation as in Figure 11.1 for coal mine waste. They stated that: 'To initiate the sequence, pyrite is oxidized directly by oxygen (a) or is dissolved and then oxidized (a'). The ferrous iron formed is oxygenated extremely slowly (b) and the resultant ferric iron is rapidly reduced by pyrite (c), releasing additional acidity and new Fe (II) to enter the cycle via (b). Once this sequence has been started, oxygen is involved only indirectly in the reoxidation of ferrous iron (b), the oxygenation of FeS_2 (a) being no longer of significance. Precipitated ferric hydroxide serves as a reservoir for soluble Fe (II) (d). If the regeneration of Fe (II) decreases, it will be replenished by dissolution of the solid $Fe(OH)_3$'. In the soil, the reaction (b) is mediated by iron bacteria, so it may not be seriously rate limiting. However, it has to be noted that for the pyrite oxidation process to operate smoothly, it is necessary to maintain a high level of ferric iron activity, and for this to be realized, the medium pH must be kept sufficiently low. In this relation, Murakami (1965) remarked that liming at the early stage of the reclamation of sulfide-containing sediments would retard the pyrite oxidation.

Jarosite is an intermediate oxidation product of ferrous sulfate in an acid medium:

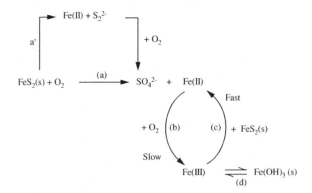

Figure 11.1
Model for the oxidation of pyrite (Source: Stumm and Morgan, 1970)

$$Fe^{2+} + SO_4^{2-} + (1/4)O_2 + (3/2)H_2O + (1/3)K^+ =$$
$$(1/3)KFe_3(SO_4)_2(OH)_6 + H^+ + (1/3)SO_4^{2-}$$

It appears in the pores and cracks as pale yellow (or straw yellow) mottles and is indicative of acid sulfate soils. It is further hydrolyzed in a less acidic medium, finally to precipitate iron as hydrated ferric oxides, so releasing more acidity.

$$2KFe_3(SO_4)_2(OH)_6 + 6H_2O = K_2SO_4 + 6Fe(OH)_3 + 3H_2SO_4$$

The acidity produced by pyrite oxidation may be neutralized if there is abundant $CaCO_3$ in the sediment or introduced in water. In some coastal areas, many fine gypsum crystals can be seen in the soil, caused by neutralization reactions. However, in most areas of the humid tropics, the carbonate content is low or nil, and the acidity remains in the soil to create a pH sometimes as low as 3. Van Breemen and Wielemaker (1974) indicate that the assemblage of original clay minerals, ferric oxides, jarosite, basic aluminum sulfate ($AlOHSO_4$), amorphous silica and sometimes gypsum, make a nearly ideal pH-stat, keeping the pH close to 3.6–3.8, in the case of acid sulfate soils in the Bangkok Plain of Thailand. They state that sediments with low amounts of smectite clays, may have an equilibrium pH close to 3.0–3.5. If we leave these low pH acid sulfate soils under the natural submergence and draining cycles, the deacidification mechanism would operate to raise the soil pH finally to 4.5–5 (see Chapter 7).

11.1.2 Reclamation and amelioration of acid sulfate soils for rice cultivation

Because of the generation of strong acidity, if acid sulfate soils are to be reclaimed for agricultural production, human intervention is necessary. As acid sulfate soils usually occur in swampy lowlands, rice is often the first crop to be considered. Various factors involved in establishing paddy fields in an acid sulfate soil area are considered next.

Rice is known to tolerate relatively high acidity. In a solution culture, only a pH below 4 affects rice adversely (Ponnamperuma et al., 1973). Under such a low pH, aluminum toxicity may be more important than the direct effect of the hydrogen ion concentration. In some studies conducted in Japan, the growth of rice seedlings began to be inhibited when the Al^{3+} ion concentration exceeded 35–40 mg kg^{-1}. Cate and Sukhai (1964) and others quoted 25 mg kg^{-1} as the threshold at which rice seedlings started to show aluminum toxicity. According to Van Breemen and Pons (1978), these toxic levels of aluminum can occur in acid sulfate soils with a pH of 3.5 or less.

Ferrous sulfate produced in quantity during the oxidation process of pyrite was considered to adversely affect the growth of rice (Kobayashi, 1939). Today, it is known that iron toxicity results from a complex nutrient imbalance (see Chapter 9), but a high concentration of ferrous ions, when coupled with high acidity and/or high salinity, may be an important element of disorders frequently found in acid sulfate soils.

Phosphorus probably is the most critical nutrient in acid sulfate soils. In addition to the inherent paucity of P in these soils, the high activity of aluminum lowers the availability of P to rice. It is so acutely deficient in some soils that no response, even to lime and N, can be expected, unless P is also applied. Kawaguchi and Kyuma (1969) stated that one of the benefits of liming and P application lies in the increased supply of N caused by the stimulation of ammonification. Matsuguchi *et al.* (1970) found that liming and phosphate application definitely increased the amount of micro-biologically fixed N. Thus in many acid sulfate soils, P is the limiting factor for the increased growth and yield of rice.

The harmful concentrations of toxic substances could be alleviated by:
- preventing the oxidation of pyrites contained in potentially acid sulfate sediments;
- leaching the substances out of the rooting zone after allowing oxidation to occur;
- deactivating aluminum by raising the pH of the medium through liming, which also reduces the concentration of ferrous iron in the soil solution (see Chapter 6).

All of these three measures have been recommended by researchers who worked on acid sulfate soils to make them suitable for rice cultivation.

The first measure is not a positive solution to the acid sulfate soil problem. It suppresses the natural ripening process of the so-called mud clay (potential acid sulfate sediments). Because of the strongly reductive environment and very unfavorable working conditions, the rice yield remains very low, and yet no positive measures for yield improvement can be taken. In addition, this 'apparently easy' solution is actually not easy, because of unpredictable climatic fluctuations. When unusually long dry spells occur, the soil surface is inevitably oxidized and strongly acidified, causing a total failure of the rice crop. Tanaka and Yoshida (1970) reported such an incident in Camarines Sur, the Philippines.

The second measure can only be used where good drainage facilities are provided. In one lysimeter experiment, conducted by Murakami (1965) in

Japan, the third crop (one crop per year) in a heavy clay soil (50% clay and 42% silt) plot that was provided with tile drainage at a depth of 80 cm gave over 4 t ha^{-1} of paddy, whereas a plot without drainage gave almost no yield through the first four crops, and only later did the yield increase significantly. In the first three years of percolation treatment, the oxidizable sulfur content decreased from the initial 192 mg to 12 mg kg^{-1} soil. This area has about 800 mm of precipitation during the fallow months, which must have contributed greatly to the removal of toxic substances through natural percolation.

The conditions under which acid sulfate soils occur in Southeast Asia must also be considered. Most of the areas are low-lying, and natural drainage is severely restricted. During the rainy season, the entire land is submerged, but during the dry season many of the areas become dry to a depth of a few decimeters. Surface water during the rainy season is mostly fresh and washes the top few surface centimeters of soil, but in the dry season, upward capillary movement of water cancels out this effect, carrying the toxic substances back to the surface.

The minimum requirement for the reclamation of acid sulfate soils is to dig open ditches for drainage, except when the first measure for preventing oxidation is taking place. Often tidal gates have to be installed in coastal areas to prevent the entry of salt-water during the dry season. This investment would certainly reinforce the washing effect during the rainy season, but it is not effective at all during the dry season unless there is an ample supply of freshwater, which is generally difficult to attain in the monsoon climate zone.

In areas with a permanently humid climate, as typically seen in Sarawak, East Malaysia, leaching first with sea or brackish water, followed by leaching with plenty of rainwater, could be an effective means of reclaiming acid sulfate soil. This method of soil amelioration, was previously adopted in Sierra Leone in West Africa with some success. Ridging or making a dense network of shallow ditches should also prove effective in such areas for upland crop cultivation, provided major drainage works are furnished. The *Surjan* system in Indonesia, with broad ridges (*guludan*) and shallow furrows (*tabukan*), has been the time-honored local practice in the utilization of acid sulfate soils.

The third measure, liming, is effective in raising the soil pH and thus inactivating toxic aluminum. However, if this measure alone is taken, the amount of lime required would easily rise to an uneconomically large figure. Van Breemen and Pons (1978) calculated that complete neutralization of the acidity generated in a 50 cm deep soil containing 3% pyrite would require 150 t of lime per hectare, even when half of the acidity is removed by leaching. Moreover, it must not be forgotten that the acidity is generated only slowly upon exposure of pyritic mud to the air during the dry season, and so the liming has to be repeated annually for many years.

Another point to be considered is that the oxidation of pyrite, the dominant form of oxidizable sulfur, is retarded by raising of the pH above 3. Therefore, if lime is applied at the beginning of reclamation, the time required to leach out the toxic products of oxidation would be prolonged.

Thus, liming should be carried out only after oxidation and leaching have been advanced to some extent. In Thailand, liming is recommended in amounts just sufficient to inactivate aluminum (Komes, 1973). Further liming to produce higher pH is thought to have the adverse effect of causing sulfate reduction in the rooting zone. It is known (van Breeman, 1975) that sulfate reduction begins significantly when the medium pH exceeds 5. Murakami (1965) however, indicates that acid sulfate soils usually do not suffer from the harmful effects of hydrogen sulfide because sulfate reduction is rather rare. He argues that in soils, ferric oxides are liberated and precipitated as a final product of pyrite oxidation, and so the *Eh* is kept relatively high–high enough to suppress sulfate reduction.

It is expected that oxidation and leaching, with or without liming, will cause a loss of soil mineral nutrients and accordingly, lower soil fertility. However, Murakami (1965) concluded from his experiments that such effects are not serious. His results revealed that:
   • All of the exchangeable cations decreased significantly, but Ca, Mg and K in the soil solution were kept at a level high enough to supply rice plants with these elements;
   • The CEC did not decrease appreciably;
   • The capacity for P absorption capacity did not increase;
   • The C/N ratio of soil organic matter fell from about 20 to 10, after three cropping seasons, indicating a maturation of humic substances in the soil.
Some studies conducted on the fertility status of acid sulfate soils are introduced next.

11.1.3 Fertility characteristics of acid sulfate soils
Table 11.1 gives the fertility-related properties of the soils in the Bangkok Plain of Thailand (Attanandana *et al.*, 1981). Even though the samples of acid sulfate soil are at different stages of maturity or amelioration, their pH is invariably <5, and even <4 for Rangsit very acid soils, as compared with the control non-acid sulfate soils with a pH > 5. Reflecting the low pH, exchangeable Al is also high for Rangsit very acid soil–higher by two orders

Table 11.1 Fertility characteristics of some representative acid sulfate soils and non-acid sulfate soils in the Bangkok Plain

Soil series (Soil suitability class)*	Sena (P-IIa)	Rangsit (PIIIa)	Rangsit very acid (PIVa)		Bangkok (P-I)	Ratchaburi (P-I)
Soil texture	HC	HC	HC	HC	HC	SiCL
pH (H$_2$O)	4.7	4.8	3.9	3.0	5.2	5.8
Organic matter, %	2.2	2.1	3.1	2.2	1.3	1.5
Available phosphorous, mg P kg^{-1}	9.6	10.5	6.1	7.0	23.6	50.8
Exch. K, cmol kg^{-1}	0.42	0.37	0.17	0.24	0.67	0.16
Exch. Na, cmol kg^{-1}	0.6	1.2	3.7	0.56	2.7	0.76
Exch. Ca, cmol kg^{-1}	13.1	10.0	2.5	1.25	10.0	7.5
Exch. Mg, cmol kg^{-1}	4.0	8.5	5.6	3.6	9.3	2.7
Exch. Al, cmol kg^{-1}	0.55	0.81	11.4	15.1	0.18	0.09
Available silica, mg SiO$_2$ kg^{-1}	101	109	36	-	141	114
CEC, cmol kg^{-1}	31.2	26.4	30.6	-	23.2	17.4
0.7nm minerals, %	55	50	65	-	25	50
1.0nm minerals, %	15	10	10	-	20	25
1.4nm minerals, %	30	40	25	-	55	25
Total sulfur, mg S kg^{-1}	1281	706	5235	-	1131	181
Water soluble sulfur, mg S kg^{-1}	440	176	400	-	520	0
Jarosite sulfur, mg S kg^{-1}	560	492	2287	-	0	0

* P—suitability for paddy rice; I–IV—very well suited through unsuited; a—acidity being the major constraint
(Source: Attanandana *et al.*, 1981)

of magnitude than the control. Available P and silica are lower for acid sulfate soils, and again are very low for Rangsit very acid soil. In spite of the poor chemical nature, acid sulfate soils still retain relatively high contents of 1.4 nm minerals (mainly high-charge smectites) and thus a high CEC. However, when combined with a very low pH this could lead to high exchangeable Al, as seen here for Rangsit very acid soil, that would cause Al toxicity to rice. Jarosite sulfur is found only in acid sulfate soils, but total sulfur, as well as water-soluble sulfur, are quite high, even in one of the control soils. This may be due to the presence of gypsum in the Bangkok series soil.

As shown here, the major fertility problems of acid sulfate soils are low pH, high exchangeable Al, low P and low silica. However, since the clay mineral composition has not yet been severely degraded, if acidity is carefully amended, rice is expected to respond well to fertilizer application. Attanandana *et al.* (1981) conducted a pot experiment using the five soils listed in Table 11.1, and showed that Sena and Rangsit soils could attain the same yield level as Bangkok and Ratchaburi soils, if properly limed and fertilized, particularly with P and K. However, rice in Rangsit very acid soil failed with Al toxicity symptoms, in spite of a preliminary application of sufficient lime. Most probably, this was due to a new generation of acidity in the rhizosphere caused by the the hydrolysis of basic aluminum, or iron, sulfates, like jarosite.

Phosphorus is the crucial nutrient in acid sulfate soils. In the above pot experiment, growth of rice in Rangsit very acid soils was almost completely inhibited in the no-P plot. Phosphorus in the soil would have been caught by active Al that was abundant in this soil. However, this may also indicate the effectiveness of phosphates as a suppressor of Al toxicity for rice. Using an X-ray microanalyzer, Attanandana *et al.* (1982) demonstrated that Al taken up by rice in a high phosphate plot was concentrated in the epidermis of the roots along with P. This suggested that phosphates applied near rice roots could be more effective in suppressing Al toxicity. Thus, Attanandana *et al.* conducted a field experiment in which they inserted mud balls mixed with phosphates into the root zone of rice seedlings and obtained certain positive results in the yield, compared with homogeneous application of the same amount of phosphates.

Pulverized rock phosphate without any processing may be effectively used as the source of P to take advantage of the strong acidity of acid sulfate soils. Depending on the origin of rock phosphates, the response of rice could vary quite widely but there will certainly be more use of this method in the tropics.

11.2 Peat soils

Peat occurs throughout the world where the balance between the addition and decomposition of organic matter favors its accumulation due to excessive water (or anaerobic conditions) and/or low temperature. Globally, peat covers more than 200 million ha of land, of which about 30 million ha are in the tropics. The proportion of peat in insular Southeast Asia is extraordinarily large. Land area of insular Southeast Asia is only 5% of the total land area of the tropics, whereas it has some 20 million ha or 2/3 of the total area of tropical peat, mostly around the Sunda Shelf (Driessen, 1978; Dent, 1986). The most widely accepted definition of peat soil is a soil having a minimum thickness of 50 cm of organic layer, which contains more than 65% of organic matter by weight.

11.2.1 Characteristics of tropical peat

In temperate and cold regions, it is usual to classify three groups of peat: low-moor peat; intermediate peat and high-moor peat. Low-moor peat is formed in depressions under submergence. As the depression is gradually filled with plant debris of aquatic vegetation (mainly reeds and sedges that are still rooted directly in the underlying mineral soil), intermediate peat begins to form from the plants rooted mainly in the low-moor peat. Thus, intermediate peat is poorer in mineral elements. By this time, the free water surface disappears due to the complete burial of the depression by peat. High-moor peat mainly constituted by sphagnum is formed on the intermediate peat. The surface of high-moor peat is raised over the previous free water surface, although the interstitial pores of the sphagnum peat are saturated with water. High-moor peat has the least mineral elements because it has no footing on underlying mineral soils.

In contrast to the dominantly herbaceous nature of temperate peat, tropical peat comprises mostly of the remnants of woody stems and branches of swamp forest vegetation, so is called woody peat. According to Driessen (1978), tropical peat may be classified into topogenous peat and ombrogenous peat. Topogenous peat formation is similar to that of low-moor or intermediate peat, beginning with the establishment of aquatic vegetation in a permanently submerged depression and ending in the complete burial of the depression with organic debris. The formation of ombrogenous peat occurs upon this topogenous peat under the closed canopy of swamp vegetation of year-round rainfall and excessive humidity. It gradually heaps up to form a dome, and at the same time extends radially until it covers the entire basin. Like high-moor peat, ombrogenous peat is poor in mineral elements because it is formed from the vegetation that relied on the limited amount of mineral nutrients released from the underlying peat mass. Thus, topogenous peat is mesotrophic, if not as fertile as eutrophic peat that occurs in some parts of the temperate region. Ombrogenous peat, especially the one in the center of a dome, is invariably oligotrophic.

Tropical peat, as observed in Sarawak and Sumatra, sometimes has a depth exceeding 10 m with 75–95% of its pore space filled with water, and a bulk density from 0.05–0.4 g cm^{-3}. It is very difficult to imagine how this thin slurry of semi-decomposed organic debris can support a swamp forest with 300 t ha^{-1} of above-ground biomass. According to carbon dating data, tropical peat is considered to have begun being deposited some 6000–6500 years ago, during the Hypsithermal period of the Post-Glacial age, at a mean rate of 2–3 mm per year. This is much faster than the rate of 0.6–1 mm per year measured for herbaceous peat in Japan.

11.2.2 Reclamation of tropical peat soils for rice cultivation

When peat soil is reclaimed, the area must be drained first, even when rice is the planned crop after reclamation. The solid fraction of the deep, loose

and fibrous, central dome peat occupies only 5–7% of its volume, while
that of the relatively densely packed peat may occupy only 25% of its total

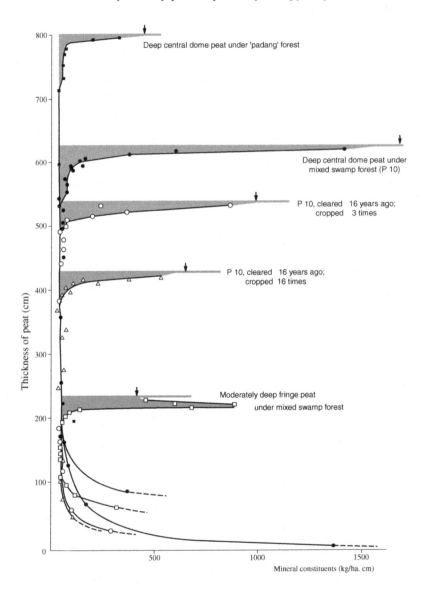

Figure 11.2
*Distribution of mineral constituents in a number of representative virgin
and reclaimed ombrogenous dome peats from West Kalimantan, Indonesia
(Source: Driessen, 1978)*

volume, with the remaining pore space being fully saturated with water. Therefore, unless drained, peat has almost no bearing capacity.

Open ditches normally one-meter deep are positioned at 20 to 40 m intervals for drainage. As drainage and the dehydration of peat proceeds, the land subside considerably. In addition, as dehydrated peat undergoes oxidative decomposition, it subsides even further. Therefore after some years, the drainage must be readjusted. A subsidence rate of 1 m y^{-1} has been recorded in the first year of reclamation of deeply drained ombrogenous woody peat.

According to Driessen (1978), tropical peat accumulates most of the available plant elements in the top 25 cm, where a dense root mat is formed (Figure 11.2). The subsurface layer commonly contains much lower amounts of mineral elements and the layers below 80 cm, which are almost the lower limit of the living root system, contain only negligible nutrients. This means that almost all the mineral elements contained in a peat soil are recycled rapidly by the vegetation. Therefore, clearing the vegetation and decomposition of the surface layer of peat inevitably lead to a rapid loss of mineral nutrients from the soil system. Driessen (1978) gave the example of a deep peat soil under virgin mixed swamp forest that contained 13250 kg of mineral constituents per hectare in its upper 80 cm layer. Of this some 10850 kg were involved in cycling, while the rest (2400 kg ha^{-1}) was in the structural component of organic materials. In the 'padang' forest, which is a poor stand that naturally occurs at the center of a peat dome and is often established after temporary agriculture is abandoned, only 5630 kg of mineral matter per hectare occurred, of which 2380 kg are integral to the peat structure. This limits the quantity of plant-available elements to a mere 3250 kg ha^{-1} for the top 80 cm.

Table 11.2, taken from Driessen (1978), shows a decline in the content of total ash, K_2O, P_2O_5, and SiO_2, after the forest vegetation is cleared, while the amount of CaO and MgO often shows an increase due to the management practice. What is more impressive is the influence of cropping on the subsidence and mineral nutrient content of peat. Figure 11.2 shows that a reclaimed area of peat subsided by some 90 cm when cropped only three times in sixteen years, while the same peat subsided by as much as 2 m in the same sixteen years when cropped every year. Figure 11.2 and Table 11.2 show that peat cropped annually suffers a much greater loss of nutrients than the same peat cropped only a few times between long fallow periods.

The N content of tropical woody peat is better compared to other nutrients, and is commonly in the order of 2000 to 4000 kgN ha^{-1} for the 0–20 cm layer. However, only less than 3% of this amount is readily available

Table 11.2 Total nutrient contents of six surface soils from lowland peat areas in Sumatra and Kalimantan

Surface soils	Nutrient contents, kg ha⁻¹					
	P_2O_5	K_2O	MgO	CaO	SiO_2	Total ash
West Kalimantan (0–20cm)						
Deep peat under light mixed swamp forest	664	119	482	444	5892	9070
Same; cleared 16 yrs ago, cropped 3 times	266	128	647	1239	1670	6570
Same; cropped 16 times	163	40	432	933	983	4340
Riau (0–25cm)						
Moder. deep peat under mixed swamp forest	217	86	685	211	14960	17500
Same; cleared 2 yrs ago, never cropped	229	50	965	1612	11870	17180
Same; cleared 30 yrs ago, perennial crops	432	74	852	3050	4400	16000

(Source: Driessen, 1978)

to plants (Driessen, 1978). Thus, N application is almost always essential for a successful cropping.

As stated above, the nutrient status of peat is generally very poor, so fertilizer application is inevitable if crops are to be cultivated with some success on reclaimed peat. Driessen's (1978) general recommendation for the application of micro and macro elements was respectively: 15 kg copper sulfate, 15 kg magnesium sulfate, 15 kg zinc sulfate, 7 kg manganese sulfate, 0.5 kg sodium molybdate and 0.5 kg borax per hectare; and 50–130 kg N, 30–70 kg P_2O_5 and 60–100 kg K_2O per hectare per year.

As peat is generally very acidic with a pH of between 3 and 5, liming is almost always essential. Chew (1971) says that 8–10 tons of ground magnesium limestone per hectare, followed by an annual application of 1 t ha⁻¹, would give a good yield for most short-term crops in acid peat soils.

Rice is considered to be a good crop in peat soils because it requires prolonged submergence of the land, which reduces the subsidence and decomposition of organic materials. Many attempts have been made to establish rice on reclaimed peat, but so far, paddy rice has not been cultivated successfully in acid tropical peat soil.

Rice becomes sterile in deep peat. According to Driessen and Suhardjo (1976), this is most probably due to imperfect photosynthesis or carbo-hydrate translocation, or a disturbance of the generative system of the rice plants, or to all three. Stagnant ground water in deep peat has a low pH and a low base status, and probably a special composition of dissolved organic compounds. These organic compounds have a strongly polyphenolic character and can cause an uncoupling of oxidative phosphorylation (Flaig, 1968), which is essential for the formation of starch in plants. Apparently,

Copper (Cu) deficiency retards the deactivation of toxic phenols, and also causes male sterility (Graham, 1975). Driessen's conclusion here is that rice sterility of rice is probably due to inhibited phosphorylation in the presence of certain phenols in a Cu deficient environment. As Cu that is applied directly to paddy fields rapidly becomes inactive, no practical solution to this problem is known.

There are about 200,000 ha of peat land in Hokkaido, Japan, of which about one-quarter has been reclaimed for agricultural use since the beginning of the 20th century. This reclamation involved three main activities:
- drainage;
- soil dressing;
- fertilization.

Drainage was carried out by digging open ditches at 540 m intervals and densely aligned tile drainage lines under the peat at 12.5 m interval. As drainage proceeded, the peat was dehydrated and decomposed. Considerable land subsidence was inevitable during the course of reclamation. The chemical composition of peat changed from the one with a high cellulose and high hemicellulose content to the one with a high lignin content in the course of maturation.

Soil dressing is thought to be an essential component of peat land reclamation. In view of the serious rice sterility problem in tropical peat areas, successful rice cultivation in Japanese peat land can be ascribed to soil dressing (see 11.2.3). About 1000 to 1500 tons of mineral soil are usually dressed per hectare, which gives a surface mineral soil layer of about 10 cm. The frozen ground and deep snow, during Hokkaido's winter would have facilitated the use of sleighs to do the laborious work of soil dressing. The effect of soil dressing, besides amending the rice sterility, may be summarized as follows:
- Bulk density is increased, making it possible to prevent the floating of rice due to gas evolution during the summer;
- Mineral elements are supplied;
- Soil temperature rises more rapidly.

Fertilization is necessary even after soil dressing. Generally peat paddy soils are poorer in P and K, and much poorer in silica than alluvial paddy soils. However, the supply of N from herbaceous peat in Japan is quite substantial, making it necessary to adjust the amount of N fertilizer to be applied. The application of magnesium oxide often appears effective for high-moor peat.

After these treatments, peat soils can be made as productive as ordinary mineral paddy soils. There has never been a rice sterility problem in

Hokkaido, and by now, the yield level of peat paddy soils is not inferior to that of mineral paddy soils.

11.2.3 A case study of tropical peat soils in Thailand and Malaysia

Kyuma *et al.* (1992) studied tropical peat soils in relation to their reclamation and agricultural utilization. Some of the more relevant data are introduced below (cf., Kyuma, 1991).

1) Contribution of peat decomposition to land subsidence

Long-term observations at the two peat experiment stations at Jalan Kebun in Selangor and Pontian in Johore, Peninsular Malaysia, show that the annual depth of land subsidence becomes almost constant at 2.5 cm, after many years of agricultural use. In order to estimate the contribution of the decomposition of peat itself to the rate of land subsidence, soil respiration was measured in the Pontian area, both in a bare field within the station, and in a nearby swamp forest at Ayer Baloi. The swamp forest was no longer intact, being located in an area installed with open drainage ditches, although it was still supporting a forest with some 290 t ha^{-1} of the above-ground biomass and 40 m tall trees.

Table 11.3 provides data on annual soil respiration at both these sites. Assuming that 40% of the total respiration at the swamp forest come from the respiration of living roots (Kira, 1976), the figure for the swamp forest was 14 t peat dry matter ha^{-1}, whereas that for the bare field in the station, amounted to 42 t peat dry matter ha^{-1}; three times that for the swamp forest. Using the bulk density values measured at the plots, the contribution of peat decomposition to the annual land subsidence rate of 2.5 cm was calculated. The results revealed that some 70% of the total subsidence was

Table 11.3   Rates of soil respiration, peat decomposition and land subsidence due to peat decomposition

	Ayer Baloi swamp forest (AB1-1)	Pontian Peat Expt. Station (PN 1-1)
Total soil respiration (t C ha^{-1} y^{-1})	11.5	21
Rate of peat decomposition (t ha^{-1} y^{-1})	14	42
Rate of land subsidence (cm y^{-1})	0.73	1.79

Notes:
1) Rate of peat decomposition in swamp forest was assumed to be 60% of total soil respiration (the rest being the respiration of living tree roots).
2) Rate of land subsidence was calculated with the measured values of bulk density, 0.192 g cm^{-3} at Ayer Bakoi and 0.235 g cm^{-3} at Pontian Station.
(Source: Kyuma, 1991)

due to the decomposition of peat itself. Clayton (1928), who studied peat in the Florida Everglades also estimated that 2/3 of the subsidence was due to decomposition. Tropical peat decomposes quite rapidly after reclamation and agricultural use. The rate of subsidence for the swamp forest in the table is apparent, as new peat formation should compensate for losses in intact forests.

2) Nutrient release from peat decomposition

The oligotrophic nature of tropical peat is well known and Table 11.4 gives additional data on this feature. Using the data in Table 11.3 and 11.4, estimates of the amounts of nutrients released annually by peat decomposition are given in Table 11.5. The extreme paucity of nutrients is evident. Considerable amounts of N may be supplied (to be taken into account in fertilizer application), but P and K levels are extremely low and hardly sufficient even for short-term crops. Micronutrients, such as Cu, Mn and Zn, would not become available to a crop, even if sufficient amounts were released, as they are chelated by the phenolic components of peat. The high values of Fe, Mn and some other elements in the field at the Pontian station, are considered to come from unnatural sources, such as the laterite paving of roads.

3) Avoiding rice sterility

Even though rice is the crop most adapted to reclaimed peat, the problem of sterility has severely limited the adoption of rice cultivation in the tropics. As reported earlier, reclaimed peat in Japan has been used for rice without encountering sterility problems, and this is thought to be due to the practice of soil dressing at the time of reclamation. In a tropical peat study project conducted in Malaysia and Thailand, Tadano et al. (1992) and Kamarudin et

Table 11.4 Nutrient contents of some tropical peats

Element		Ayer Baloi swamp forest (AB 1-1)				Pontian Peat Expt. Station (PN 1-1)			
		1st layer	2nd layer	3rd layer	4th layer	1st layer	2nd layer	3rd layer	4th layer
N	(%)	1.355	0.656	0.687	0.742	1.201	0.506	0.537	0.548
P	(%)	0.014	0.009	0.009	0.007	0.011	0.007	0.006	0.006
K	(%)	0.010	0.010	0.013	0.009	0.007	0.008	0.008	0.007
Ca	(%)	0.166	0.047	0.051	0.044	2.522	0.317	0.191	0.135
Mg	(%)	0.044	0.051	0.048	0.049	0.057	0.058	0.035	0.032
Mn	(mg kg^{-1})	5.1	3.6	2.5	2.8	31.8	10.6	9.6	11.9
Fe	(mg kg^{-1})	1034	673	346	295	2560	485	252	324
Cu	(mg kg^{-1})	1.6	1.5	1.4	1.8	100.3	6.5	3.8	4.5
Zn	(mg kg^{-1})	5.9	3.9	3.0	3.1	33.4	7.7	4.5	3.6

(Source: Kyuma, 1991)

Table 11.5 Rate of nutrient release from peat decomposition (kg ha^{-1} y^{-1})

Element	Ayer Baloi swamp forest (AB1-1)	Pontian Peat Expt. Station (PN 1-1)
N	189.7	504.4
P	2.0	4.6
K	1.4	3.1
Ca	23.2	1059
Mg	6.2	24.0
Mn	0.1	1.3
Fe	14.5	107.5
Cu	0.0	4.2
Zn	0.1	1.4

(Source: Kyuma, 1991)

al. (1992) clarified, both in the field and pot experiments, that Cu deficiency and polyphenols certainly inhibit rice, but are not necessarily the decisive factors. Krisornpornsan *et al.* (1992) conducted a pot experiment with rice using peat from the Bacho swamp, Narathiwat, Thailand. The result of the experiment, as in Table 11.6, endorsed, to some extent, the findings of Tadano *et al.* (1992). Peat alone and peat with Cu did not produce any grain, but when liming corrected the acidity, the peat yielded some grain even without Cu application and produced a fair yield of grain with Cu. When the soil was dressed, either on the surface or mixed with peat, the grain yield was increased, and further additions of Cu had no more effect.

In field conditions, it was much more difficult to grow rice in peat in Narathiwat Province, Thailand, because of sterility and the very precarious water condition in the region. Floods occurred during the rainy months and drought during the subsequent drier months. However, Attanandana *et al.* (1999) finally confirmed the results of the pot experiment in a field trial, showing that both liming and Cu application are essential for harvesting a reasonable yield of rice in peat soils.

11.3 Saline and sodic (or alkali) soils

11.3.1 Definition of salt-affected soils

Saline soils are those with an electric conductivity of saturation extract (ECe) equal to or greater than 4 dS m^{-1}, and an exchangeable sodium percentage (ESP) less than 15%. The pH of the soil is below 8.5[1].

Sodic soils are those with an ESP greater than or equal to 15% and an ECe lower than 4 dS m^{-1}. The pH of these soils exceeds 8.5 due to the frequent presence of sodium carbonate as free salt or the hydrolysis of soil

Table 11.6 Effect of soil dressing, liming and copper application on rice grown in peat in pot experiment

Treatment	Straw (g pot^{-1})		Total grains (g pot^{-1})		Unfilled grains (%)		Filled grains (g pot^{-1})	
P	0.95	b	0.00		0.00		0.00	
P + Cu	2.49	d	0.00		0.00		0.00	
P + S (7.5cm) surface-placed	92.94	bc	105.39	c	3.40	bc	101.81	cd
P + S (7.5cm) surface-placed + Cu	113.14	ab	105.06	c	2.20	c	102.75	cd
P + S (15cm) surface-placed	124.01	ab	119.98	bc	5.60	ab	113.26	bc
P + S (15cm) surface-placed + Cu	162.26	a	127.76	b	4.40	bc	122.14	b
P + S (7.5cm) mixed	97.00	abc	103.40	c	3.70	bc	99.57	cd
P + A (7.5cm) mixed + Cu	96.92	abc	108.97	c	4.20	bc	104.39	cd
S	163.73	a	160.76	a	4.20	bc	154.01	a
S + Cu	151.46	ab	159.77	a	3.00	bc	154.98	a
P + L	38.86	cd	45.10	d	9.10	a	41.00	e
P + L + Cu	119.20	ab	112.00	bc	4.40	bc	107.07	cd

P: peat, S: soil, L: lime, Cu: copper
Notes: There is no significant difference between figures with the same alphabetical letters.
(Source: Krisornpornsan *et al.*, 1992)

exchangeable Na. The term 'alkali soils' is no longer used in the USA and some other countries.

There are also soils with ECe $\geq$ 4 dS m^{-1} and ESP $\geq$ 15%, which are usually called saline-alkali (sodic) soils. However, some authors are critical of this term because these soils could be grouped with either of the saline or sodic soils depending on their behavior.

In recent years, a new criterion was introduced to distinguish sodic soils from saline soils. The sodium adsorption ratio (SAR) measured using the soil saturation extract replaces the ESP. The SAR is defined as follows:

$$SAR = Na^+ / \sqrt{Ca^{2+} + Mg^{2+}}$$

where Na$^+$, Ca^{2+} and Mg^{2+} are determined in mmol litre^{-1}, and so the dimension of the SAR is (mmol litre^{-1})$^{0.5}$. The SAR can be theoretically related to ESP by the following equation[2]:

$$ESP/(100 - ESP) = 0.5 \times SAR/31.6 = 0.0158 \ SAR$$

A statistically obtained coefficient from many determinations was 0.0148, which is fairly close to the one derived above. According to the above formula, SAR $\geq$ 13 corresponds to ESP $\geq$ 15 (to be exact, ESP 17).

The SAR is preferred to ESP as the criterion, because of its ease of determination compared to measuring the ESP.

11.3.2 Saline soils

Because of the high osmotic tension of soil solutions, saline soils are not suitable for crop growth. As saline soils occur normally in arid to semi-arid climatic regions, desalinization by leaching is not readily practiced unless there is a source of good quality irrigation water. However, recently much salt-affected land has been irrigated and numerous studies have been conducted of the amount and method of water application. The water required to leach out salts from the rooting depth of crops depends on the texture and permeability of the soil. For example, the quantity of water required to leach clay loam soils is 3–4 times that of silt loam soils even when the latter contains more salts (Hulsbos, 1963).

As a leaching method, alternate submergence and drying is often practiced to increase the efficiency of desalinization. During submergence, rice may be cropped to take advantage of the crop's adaptation to excessive water. Rice also seems to be tolerant of highly saline conditions. Normally rice can endure salinity up to 4 dS m^{-1} but some studies indicate that rice can tolerate much higher salinity. Of course, the tolerance of rice varies, depending on the variety and the stage of growth. During the seedling and heading stages, rice is more sensitive to salt. Even so, Yoneda (1964) states that rice grown in the Kojima polder of Okayama Prefecture in Japan, could tolerate salinity up to 7 dS m^{-1}. Van Alphen's report (1985) sets the limit at an even higher level of 20–25 dS m^{-1}.

Thus, the high tolerance of rice to both excessive water and salinity is considered to make it the most suitable crop to grow during the course of ameliorating saline soils by irrigation. According to Bhumbla and Abrol (1978), even in soils with a low infiltration rate, the percolation of accumulated water during a rice season amounts to 100–300 mm, which helps leach salts to a level low enough for subsequent crops; for which legumes are often recommended.

11.3.3 Sodic Soils

Sodic soils are characterized by two adverse conditions for crops. One is a highly alkaline reaction, and the other is a poor physical condition due to a high percentage of Na in the exchange complex. Often sodium carbonate is the main soluble salt and this gives a high pH, sometimes as high as 10.5, because of hydrolysis. The high pH causes nutrient imbalance and low availability of certain micronutrients.

Ready dispersion of soil clays due to high ESP is the cause of structure deterioration, and consequently, low available water holding capacity and reduced permeability. These conditions are fatal for most upland crops.

For the reclamation of sodic soils, gypsum ($CaSO_4 \cdot 2H_2O$) is most commonly used under irrigation to effect the exchange of Na for Ca. Rice seems to be ideal for the ameliorative stage of sodic soils, because it can withstand excessive water and its requirement for soil physical properties is minimal. As discussed earlier, the highly alkaline pH of sodic soil may not be a problem when the soil is submerged, as alkaline soil pH is lowered by carbon dioxide evolved in the course of reduction. Also, rice is very tolerant of exchangeable sodium. In one experiment, it was shown that at ESP 50, the rice yield was nearly unaffected, whereas the wheat yield, a crop considered quite tolerant to sodicity, was reduced by about 50% (Bhumbla and Abrol, 1978).

The high partial pressure of CO_2 in submerged paddy fields also facilitates mobilization of native calcium carbonate (which is almost invariably present in sodic soils) and accelerates the exchange of Na for Ca (Chhabra and Abrol, 1977).

The very high pH of sodic soils seems to be tolerated by rice, but it develops certain fertility problems like ammonia volatilization loss and zinc deficiency (see Chapters 8 and 9). Sometimes even Fe deficiency is reported. However, if amended properly for micronutrients, particularly zinc, and applied with gypsum and N, rice yields as high as 9 t ha^{-1} may be attained even in a highly sodic soil (Bhumbla and Abrol, 1978).

11.4 Soils polluted with hazardous elements

11.4.1 Soil pollution with hazardous elements in Japan
Soil is one of the most important environmental resources of the human race, along with atmosphere and water. It is considered to be part of the global commons that support not only human beings, but also all the plants and animals on earth. In spite of this importance, it has often been maltreated, resulting in wide areas of non-productive soils due to degradation and pollution.

Soil pollution with hazardous elements would have occurred from the time humankind started to use various heavy metals, such as copper and silver. However in many countries, large-scale pollution became obvious only after they started modernization or industrialization.

During the 1880s, the operation of the Ashio copper mine, located in the upper reaches of the Watarase River north of Tokyo, aroused the first serious public concern over environmental pollution in Japan. The pollution was caused directly by acid waste-water and arsenic-containing SO_2 gas from the smelter. Indirectly, the felling of mountain forest for fuelwood weakened the resistance of the land to heavy rains, and the runoff washed out a tremendous amount of mine waste into the river. Extensive paddy lands in the middle and lower reaches of the river were affected by frequent floods that carried noxious minerals (Iimura, 1991). Apart from crop failure, human and animal health hazards were also reported and the lands were left barren. Although the mine was closed in 1973, even today quite a large area of paddy fields within the watershed is still heavily polluted with copper and cannot be cropped without costly remediation measures.

Another example of soil pollution that generated serious human health problems in Japan, occurred in the lower-reaches of the Jintsu River in Toyama Prefecture, along the Japan Sea coast. People in the area suffered from what was called *itai-itai* disease, which has symptoms similar to those of rheumatism or osteomalacia (softening of the bones). The name of the disease literally means severe pain. Cases were recorded in the 1920s, but public concern was aroused only in the 1950s. After many years of investigation, the cause of the disease was finally traced to cadmium (Cd) in 1968. A mine in the upper reaches of the Jintsu River, had been producing zinc and lead. Some of the mine spoils, containing Cd as a contaminant, were carried down by the river, and deposited in paddy fields as the water was introduced for irrigation. Rice grown in the fields contained a high level of Cd, often >1 mg kg^{-1} in brown rice, so that long-term ingestion of polluted rice caused health problems. Cadmium tends to accumulate in the kidney cortex as a prodromal stage of the *itai-itai* disease (Iimura, 1991; Asami, 1991).

There have been many other incidences of soil pollution with hazardous elements in Japan, mainly Cd, Cu and As. Accordingly in 1970, the Japanese government enacted a law to prevent the pollution of agricultural soils. The law designated these three elements as the specific soil pollutants for which the governor of each prefecture must take necessary measures to prevent crop contamination. The measures consist mainly of soil dressing to cover the polluted soil, but in the case of severe pollution removal of polluted soil and dressing with unpolluted soil is conducted. The criteria for judging the pollution are as follows:

• Cadmium (Cd) >1.0 mg kg^{-1} of total Cd contained in brown rice

- Copper (Cu) >125 mg kg^{-1} of 0.1N HCl soluble Cu contained in paddy soils
- Arsenic (As) >15 mg kg^{-1} of 1N HCl soluble As contained in paddy soils

As is evident from these criteria, the law only targets paddy soils. As of 2001, 143 localities with 8410 ha exceeded the criteria for one or more of the three elements, of which 93 localities with 6616 ha were polluted with Cd (Asami, 2002). Also, a nation-wide survey carried out in relation to the law disclosed that 9.5% of paddy fields, 3.2% of upland fields, and 7.5% of orchards were more or less polluted with various hazardous elements (Min. of Agric., 1974).

Table 11.7 compiled by Iimura (1991) gives the natural abundance levels of various hazardous elements in the earth crust, shales, river water, Japanese soils as a whole, Japanese paddy soils, and Japanese brown rice. The natural abundance of these hazardous elements in soils is comparable with those of soils around the world (Bowen, 1979), except for mercury and arsenic. This indicates that unless polluted Japanese soils are not particularly high in most hazardous elements, and no chronic pollution is yet evident. However, the content of Cd in brown rice (0.09 µg g^{-1} or mg kg^{-1}), is already quite high compared with other countries, as shown in Figure 11.3. It is probably due to this fact that the daily intake of Cd by Japanese people is much higher (ca. 30–70 µg d^{-1}) even in unpolluted areas, than people (<30µg d^{-1}) who live in other developed countries (Asami, 1991).

As shown here, the pollution of paddy soils with hazardous elements can seriously affect not only the productivity of rice, but also human health, through the uptake of those elements by rice plants. In the following section, the behavior of the respective elements in paddy soils will be considered.

11.4.2 Behavior of hazardous elements in paddy soils

1) Cadmium (Cd)

The Japanese law for the prevention of soil pollution sets the permitted level of Cd in brown rice but not in the soil. This is different from the two other designated elements in the law and is considered to have emerged from the legislators' strong concern about human health. Whatever the reason may be, one problem is that no correlation has yet been established between the amount in rice and the amount in soil. Also, the amount in rice fluctuates widely even when grown in the same soil from one year to another.

Table 11.7 Natural abundance of hazardous elements (mg kg⁻¹)

Atomic number	Element	Earth crust	Shales	River water (µg kg⁻¹)	Soil (all inclusive) mean	(range)	n*	Paddy soil mean	(range)	n	Brown rice mean	(range)	n
24	Cr	100	90	1	50	(3.4–810)	370	64	(16–337)	190	-	-	-
27	Co	25	19	0.2	10	(1.3–116)	437	9	(2.4–23.5)	169	-	-	-
28	Ni	75	68	0.3	28	(2–660)	767	39	(9–412)	379	0.19	(.07–.25)	30
29	Cu	55	45	7	34	(4.4–176)	838	32	(11–120)	408	2.9	(n.d.–15.5)	2445
30	Zn	70	95	20	86	(9.9–622)	911	99	(13–258)	408	19	(n.d.–50.8)	2443
33	As	1.8	13	2	11	(0.4–70)	358	9	(1.2–38.2)	97	0.16	(n.d.–1.76)	2712
42	Mo	1.5	2.6	1	2.6	(0.2–11.3)	115	-	-	-	-	-	-
48	Cd	0.2	0.3	0.1	0.44	(.03–53)	465	0.45	(.12–1.41)	147	0.09	(n.d.–.88)	8163
80	Hg	0.08	0.4	0.07	0.28	(n.d. –5.36)	683	0.32	(n.d.–2.90)	292	0.013	(n.d.–.17)	318
82	Pb	12.5	20	3	29	(5–189)	773	29	(6–189)	407	0.19	(n.d.–2.11)	2616

* 'n' denotes number of samples analyzed. '-' denotes no available data. 'n.d.' denotes 'not determinable'.
(Source: Iimura, 1991, partly modified: For original sources of data, refer to Iimura, 1991.)

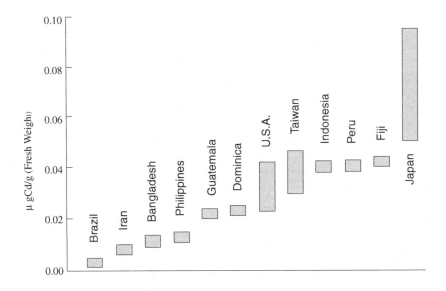

Figure 11.3
Comparison of Cd contents in rice grains produced in various countries
(Source: Asami, 1991)

In a study conducted at the Hyogo Agricultural Experiment Station, it was found that, although there was no significant correlation between the amounts of Cd in rice and those in soil analyzed with 0.1 N HCl extract, a fairly significant correlation was established between the Cd content of brown rice and 0.01 N HCl soluble Cd in the soil. The latter is also highly negatively correlated with soil pH and positively correlated with soil Eh, suggesting that Cd uptake by rice would decrease as the soil pH is raised, and the soil Eh is lowered. The Cd content of brown rice clearly increased when water saving cultivation methods, such as intermittent irrigation, were adopted. Also, the annual fluctuation of Cd in rice was caused by the fluctuation of the soil Eh during the cropping season (Jikihara, 1991). It is known that Cd becomes insoluble under reduced conditions, as it precipitates as CdS, reacting with sulfide formed by sulfate reduction.

In further studies, the same group confirmed the following (Kusaka *et al.*, 1979; Yoshikawa *et al.*, 1978; Yoshikawa *et al.*, 1979):

• Liming materials, like Ca-silicate, fused phosphate, etc., were effective in suppressing Cd uptake by rice. Fused phosphate alone and Ca-silicate + rice straw were particularly effective;

- More than 25 cm of soil dressing (placed on top of the polluted soil) was necessary to suppress Cd uptake to a sufficiently low level;
- Manganese (Mn), iron (Fe) and sulfur (S) appeared effective in suppressing Cd uptake by rice. Mn and Fe, when ionized in reduced soils, seemed to act antagonistically to Cd when being taken up by rice. Sulfur could supplement the sulfide source of soil to precipitate Cd as CdS. Part of the effect of fused phosphate stated above could have been due to Mn;
- Water management is essential in controlling Cd uptake by rice. As the greater part of Cd in rice is absorbed after its primordial initiation stage, paddy fields should be kept submerged under reduced conditions throughout that stage, until forty days after heading.

Jikihara (1991) pointed out that some of these counter-measures to Cd uptake by rice would conflict with, for example, the measures taken for arsenic. As stated below, arsenic becomes more toxic in its reduced form and tends to be more activated in a neutral to weakly alkaline pH. The way of water management needed to maintain submergence until shortly before harvest would conflict with the need for good field trafficability when using heavy harvesting machinery. Also soil dressing has many problems. It is expensive, and finding unpolluted soil for dressing in the vicinity can be difficult. Even after dressing, unproductive soil dressed on paddy fields requires care to maintain its fertility, and yet the crop performance is very often unsatisfactory.

The current criterion of 1 mg kg^{-1} Cd in brown rice is criticized as having no scientific basis. As already stated, the daily Cd intake of the Japanese people is much higher than that of people in other countries. Presently, the Codex Committee on Food Additives and Contaminants of FAO/WHO studies the international standard to be set for Cd in various foods. It set a tentative criterion for cereals, including rice, at 0.2 mg kg^{-1} in 2001, although the standard has not been finally adopted (Asami, 2002). However, if this 0.2 mg kg^{-1} were adopted, some of the rice produced in Japan would be labeled as unfit for human consumption. Much effort has been made to establish physical, chemical or biological measures for the remediation of Cd-polluted soils, but no suitable methods other than soil dressing have so far been developed.

2) Copper (Cu)

Much of the classic heavy metal soil pollution in Japan has been due to Cu–starting from the case of the Ashio mine. Copper is an essential microelement for plants, but an excess can easily cause toxicity. The natural level of Cu in Japanese paddy soils is 32.8 mg kg^{-1} (see Table 11.7), while the criterion set by the Japanese law for prevention of soil pollution is 125 mg kg^{-1} as 0.1 N HCl soluble Cu. The area of Cu-polluted paddy soil in Japan totals 1403 ha in thirty-six localities. However, copper toxicity problems occur more frequently for upland crops than paddy rice.

Rice shows a specific chlorosis about one month after transplanting in polluted soils due to iron deficiency induced by Cu toxicity. This symptom disappears as the rice grows, but in heavily polluted paddy soils, often around the inlet for irrigation water, rice dies. A regression equation for brown rice yield (Y in kg 10a^{-1}) and soil Cu concentration (X in mg kg^{-1}) was obtained in the Ariga mine area of Hyogo Prefecture (to the north of Kobe) as follows:

$$Y = -0.3X + 433 \quad (r = -0.866, \text{ significant at the } 0.1\% \text{ level})$$

According to this equation, when the soil Cu concentration is 125 mg kg^{-1}, the yield reduction is about 9%, indicating the adequacy of the threshold concentration set for Cu pollution (Jikihara et al., 1983).

There are several possible counter-measures to Cu pollution including making Cu insoluble by raising the soil pH with Ca-silicate, soil removal and dressing, and turning-over of the subsoil. Application of Ca-silicate was as effective as other measures for a few years, but it tended to lose effectiveness over time as Cu became reactivated. Consequently, the turning-over of the subsoil of up to 40 cm depth was finally adopted in the Ariga mine area as the counter-measure (Jikihara et al., 1983).

3) Arsenic (As)

The natural abundance of As in the world's soils is 6 mg kg^{-1} (Bowen, 1979), while that in Japanese soils is 11 (see Table 11.7). According to the Environment Protection Agency of Japan, the mean content of As in paddy soils and orchard soils is 8.3 mg kg^{-1} and 16.2 mg kg^{-1}, respectively. Paddy soils polluted with As, as designated by the criterion of 15 mg kg^{-1} of 1 N HCl soluble As, occur in fourteen localities with a total area of 391 ha.

The high content of As in orchard soils may result from the past use of lead-hydrogen-arsenate as an insecticide. Due to its highly residual nature, this chemical was banned many years ago. However, some of the paddy fields converted from orchards are still suffering from disorders due to As.

Arsenic has two valence states—arsenate AsO_4^{3-} and arsenite AsO_3^{3-}, of which the latter is more water soluble and more toxic. This makes As a more hazardous element in paddy soils than in upland soils, because submergence reduces arsenate to arsenite. This is the reason why paddy fields converted from orchards tend to suffer from As toxicity.

Arsenic occurs together with many heavy metal ores, such as silver, copper and zinc, as is the case with cadimium. Therefore, there are many cases of multiple pollution with As and Cd. This increases the complexity of the problems. Cadmium can be suppressed by reducing soils with submergence, whereas As becomes more soluble and more toxic under submerged conditions. Moreover, Cd is made less soluble by raising the pH above neutrality, whereas As has a higher solubility in the alkaline pH range. Thus, careful management of water and soil additives becomes necessary for incidents of multiple pollution with Cd and As.

4) Accumulation of hazardous elements in the soil

Most heavy metals tend to accumulate in the soil because of their strong tendency to undergo 'specific adsorption' onto the mineral and organic components of the soil. A general order of strength of specific adsorption for divalent cations is Pb, Cu > Zn > Co, Cd, all of which are preferentially adsorbed by the soil exchange complex relative to Ca. At a certain level of accumulation, they tend to be more readily absorbed by plants, as with Cd. Therefore, care must be taken not to allow hazardous elements to be introduced into the soil.

If soil accumulation is taken into consideration, chemical fertilizers can be an important source of hazardous elements in cultivated soils. Cadmium is almost invariably contained in commercial phosphate fertilizers at levels higher than 100 mg kg^{-1}. Some of the micronutrient fertilizers for iron, manganese and zinc often contain a high level of As and Pb. There is a tendency in many countries for establishing regulations on the content of hazardous elements in mineral fertilizers. The Fertilizer Regulation Act of Japan determines the maximum allowable contents of As, Cd, Ni, Cr, Ti and others in different kinds of fertilizers.

The same applies to organic fertilizers. In recent years, sewage sludge has been produced in large quantities from waste-water treatment plants. As the sludge contains readily available N and P, it seems appropriate to return these nutrients to cultivated soils. Sometimes sludge is processed into compost or farmyard manure with some additives and sold as organic fertilizers.

Here too, a problem arising from the use of sewage sludge is the accumulation of various hazardous elements in the soil. Table 11.8 gives the mean contents of hazardous elements in sewage sludge, night soil sludge and soils, as compiled by the Environment Protection Agency of

Table 11.8 Mean concentration of hazardous elements in sludge and soil (mg kg⁻¹ dry matter)

	Zn	Cu	Cr	Pb	Ni	As	Cd	Hg
A: Sewage sludge	961	173	49.0	48.2	37.4	6.90	2.26	1.02
B: Night soil sludge	740	121	18.1	12.3	18.7	2.86	2.14	1.10
C: Soil	54.9	24.8	25.7	17.1	18.6	6.82	0.33	0.20
A/C	17.5	7.0	1.9	2.8	2.0	1.0	6.8	5.0

Notes: A and B cited from EPA, Japan, 1983; C cited from EPA, Japan, 1984. Hg in C cited from Gotoh *et al.*, 1978. (Source: Kuboi, 1991)

Japan (1983, 1984). In many cases, the contents of zinc (Zn), Cd, Cu and mercury (Hg), are higher in sewage sludge than in the soil (Kuboi, 1991). Therefore, if sludge is repeatedly applied to the soil, these and other hazardous elements are inevitably accumulated.

In Japan, an administrative guideline was officially announced in 1984 to protect cultivated soils from the accumulation of hazardous elements. The guideline sets the upper limit of total Zn content allowable in the soil at 120 mg kg⁻¹. Thus, when the total content of Zn in the soil is 120 mg kg⁻¹ or higher, further application of organic amendments, such as sludge preparations, should be stopped. Although the validity of the standard may be argued, it is useful as a guideline for the public to judge the adequacy of the use of organic amendments.

Generally, organic amendments have been used more in upland soils than in paddy soils. However, for those farms which grow paddy rice organically, the above considerations may be relevant.

Notes

(1) Most saline soils contain free $CaCO_3$ and this controls soil pH below 8.5 in equilibrium with the atmospheric CO_2, as already explained in Chapter 6.

(2) If it is assumed that Ca^{2+} predominates among the divalent cations in a sodic soil, the exchange equilibrium between Na^+ and Ca^{2+} in the soil and soil solution is expressed by Gapon's equation:

$$X_{Na+} / X_{Ca2+} = K_G M_{Na+} / \sqrt{M_{Ca2+}}$$

where X_{Na+} and X_{Ca2+} are exchangeable cations in cmol kg⁻¹ soil, M_{Na+} and M_{Ca2+} are dissolved cations in mol litre⁻¹ in the equilibrium solution, and

K_G (mol litre^{-1})$^{-0.5}$ is the Gapon's constant, which is equal to 0.5 for a wide range of equivalent fractions of Na$^+$.

As ESP equals to $100 \times X_{Na+}/(X_{Na+} + X_{Ca2+})$:

$$X_{Na+}/X_{Ca2+} = ESP/(100 - ESP)$$

Thus:

$$ESP / (100 - ESP) = 0.5 \times M_{Na+} / \sqrt{M_{Ca2+}} \qquad (i)$$

The SAR is, as stated above, defined originally by millimolar concentration; so is denoted as 'm'. If the SAR is redefined by molar concentration, 'M', it is rewritten as follows:

$$SAR = m_{Na+} / \sqrt{m_{Ca2+} + m_{Mg2+}}$$
$$= \sqrt{1000} \times M_{Na+} / \sqrt{M_{Ca2+} + M_{Mg2+}}$$
$$= 31.6 \times M_{Na+} / \sqrt{M_{Ca2+} + M_{Mg2+}}$$

If Ca^{2+} is again presumed to predominate, then from equation (i):

$$ESP/(100 - ESP) = 0.5 \times SAR/31.6 = 0.0158 \ SAR$$

References

Asami, T. 1991. Future problems concerning soil pollution with hazardous elements. In Koyama, T. (Ed.) *Soil Pollution with Hazardous Elements–State of Affairs, Counter-measures, and Prospects*, pp. 113–135. Hakuyusha, Tokyo. (In Japanese).

Asami, T. 2002. For the solution of food and agricultural problems in Japan. *Man & Environ.*, 28: 28–37. (In Japanese).

Attanandana, T., Vacharotayan, S. and Kyuma, K. 1981. Chemical characteristics and fertility status of acid sulfate soils of Thailand. *Proc. of the Bangkok Symp. on Acid Sulfate Soils*, pp. 137–156.

Attanandana, T., Vacharotayan, S. and Kyuma, K. 1982. Fertility problems of acid sulfate soils of Thailand. *First Int'l Symp. on Soil, Geology and Landforms–Impact on Land Use Planning in Develop. Countries, Bangkok*, pp. C1-1–C1-9.

Attanandana, T., Chakranon, B., Kyuma, K. and Moncharoen, P. 1999. Improvement of a peat soil for rice cultivation in Thailand. *Jpn. J. Trop. Agric.*, 43: 91–96.

Bhumbla, D.R. and Abrol, I.P. 1978. Saline and sodic soils. In IRRI (Ed.) *Soils and Rice*, pp. 719–738. IRRI, Los Baños, Philippines.

Bowen, H.J.M. 1979. *Environmental Chemistry of the Elements* (Translated by Asami, T. and Chino, M. 1983. Hakuyusha, Tokyo. In Japanese).

Cate, R.B. and Sukhai, A.P. 1964. A study of aluminum in rice soils. *Soil Sci.*, 98: 85–93.

Chew, W.Y. 1971. Yield and growth of some leguminous and root crops on acid peat to magnesium lime. *Malays. Agric. J.* 48: 142–158.

Chhabra, R. and Abrol, I.P. 1977. Reclaiming effect of rice grown in sodic soils. *Soil Sci.*, 124: 49–55.

Clayton, B.S. 1928. Subsidence of Florida peat soil. *Trans. Int'l Soc. Soil Sci., Comm. 6B, Zurich*, pp. 840–843 (Cited from Coulter, 1957).

Coulter, J.K. 1957. Development of the peat soils of Malaya. *Malay. Agric. J.*, 33: 63–81.

Dent, F.J. 1986. Southeast Asian coastal peats and their use–An overview. In *Classification and Utilization of Peat Land: Proc. of the 2ⁿᵈ Int'l Soil Management Workshop, Thailand/Malaysia*, pp. 27–54.

Driessen, P.M. 1978. Peat soils. In IRRI (Ed.) *Soils and Rice*, pp. 763–779. IRRI, Los Baños, Philippines.

Driessen, P.M. and Suhardjo,H. 1976. On the defective grain formation of sawah rice on peat. *Soil Res. Inst. Bull., Bogor, Indonesia*, 31: 56–73.

Environment Protection Agency, Japanese Government 1983. (Cited from Kuboi, 1991. In Japanese).

Environment Protection Agency, Japanese Government 1984. (Cited from Kuboi, 1991. In Japanese).

Flaig, W. 1968. Effect of humic substances on plant metabolism. *Proc. 2ⁿᵈ Int'l Peat Congr., Leningrad.* (Cited from Driessen, 1978).

Gotoh, S., Tokudome, S. and Koga, H. 1978. *Soil Sci. & Plant Nutr.*, 24:391–406. (Cited from Kuboi, 1991. In Japanese).

Graham, R.D. 1975. Male sterility in wheat plants deficient in copper. *Nature*, 254: 515. (Cited from Driessen, 1978).

Hulsbos, W.C. 1963. Leaching of saline soils. In P.J.Dielman (Ed.) *Reclamation of Salt Affected Soils in Iraq*, pp. 36–47. ILRI Pub. 11, Wageningen. (Cited from Bhumbla and Abrol, 1978).

Iimura, K. 1991. History of soil pollution with heavy metals. In Koyama, T. (Ed.) *Soil Pollution with Hazardous Elements–State of Affairs, Counter-measures, and Prospects*, pp. 7–42. Hakuyusha, Tokyo. (In Japanese).

Jikihara, T. 1991. Countermeasures to soil pollution with heavy metals and their problems. In Koyama, T. (Ed.) *Soil Pollution with Hazardous Elements–State of Affairs, Counter-measures, and Prospects*, pp. 59–88. Hakuyusha, Tokyo. (In Japanese).

Jikihara, T., Yoshikawa, T., Tanaka, H. and Kusuka, S. 1983. *Res. Rep. Hyogo Agric. Expt. Sta.*, 31: 49–52. (Cited from Jikihara. 1991. In Japanese).

Kamarudin, A., Zahari, A.B. and Tadano, T. 1992. The effect of liming and micronutrient application on the growth of crop plants and occurrence of sterility in a Malaysian peat soil. In Kyuma, K., Vijarnsorn, P. and Zakaria, A. (Eds.) *Coastal Lowland Ecosystems in Southern Thailand and Malaysia*, pp.380–397, Kyoto Univ., Kyoto.

Kawaguchi, K. and Kyuma, K. 1969. *Lowland rice soils in Thailand*. Center for Southeast Asian Studies, Kyoto University, Kyoto.

Kira, T. 1976. *Terrestrial Ecosystems–An Introduction*. Ecology Course 2. Kyoritsu-Shuppan, Tokyo. (In Japanese).

Kobayashi, T. 1939. Studies on amelioration of problem soils in polder lands. *Spec. Rep. Ibaraki Agric. Expt. Sta.*, 3: 1–47. (In Japanese).

Komes, A. 1973. *The Reclamation of Some Problem Soils in Thailand. Soils of the ASPAC Region; Part 5. Thailand*. Tech. Bull. No. 14, Food and Fertilizer Technology Center, Taipei.

Krisornpornsan, B., Attanandana, T. and Kyuma, K. 1992. Effect of soil dressing on rice in tropical peat soils. In Kyuma, K., Vijarnsorn, P. and Zakaria, A. (Eds.) *Coastal Lowland Ecosystems in Southern Thailand and Malaysia*, pp.312–317, Kyoto Univ., Kyoto.

Kuboi, T. 1991. Heavy metal problems in the recycling of sludge to soil. In Koyama, T. (Ed.) *Soil Pollution with Hazardous Elements–State of Affairs, Counter-measures, and Prospects*, pp. 89–112. Hakuyusha, Tokyo. (In Japanese).

Kusaka, S., Ohtani, Y., Imai, T. and Jikihara, T. 1979. *Chugoku District Agric. Res.*, 44: 26–29. (Cited from Jikihara, 1991. In Japanese).

Kyuma, K. 1991. Problems related to reclamation and development of swampy lowlands in the tropics. *Jpn. J. Soil Physic. Cond. Plant Growth*, No. 63: 43–49. (In Japanese).

Kyuma, K., Vijarnsorn, P. and Zakaria, A. (Eds.) 1992. *Coastal Lowland Ecosystems in Southern Thailand and Malaysia*, Kyoto Univ., Kyoto.

Matsuguchi, T., Tangcham, B. and Pakiyuth, S. 1970. Nitrogen-fixing microflora and its activity in paddy soil of Thailand. *Proc. 1st Asean Soil Conf., Bangkok, Thailand.*

Ministry of Agriculture, Japanese Gov't 1974. Survey Results of the Soil Pollution Prevention Project in 1973. (Cited from Asami, 1991. In Japanese).

Murakami, H. 1965. *Studies on the Characteristics of Acid Sulfate Soils and Their Amelioration.* Doctoral Dissertation, KyotoUniversity (In Japanese).

Nippon Dojyo Kyokai (Jpn. Soil Assoc.) 1984. *Report of Survey and Analysis of Natural Abundance of Heavy Metals* (A Study Entrusted by Environment Protection Agency). (Cited from Kuboi, 1991. In Japanese).

Ponnamperuma, F.N., Attanandana, T. and Beye, G. 1973. Amelioration of three acid sulfate soils for lowland rice. *Proc. Int'l Symp. Acid Sulfate Soils,* vol. II: 391–406. ILRI Publ. 18, Wageningen.

Stumm, W. and Morgan, J.J. 1970. *Aquatic Chemistry: An Introduction EmphasizingChemical Equilibria in Natural Waters.* John Wiley & Sons, N.Y.

Suzuki, S., Djuangshi, N., Hyodo, K. and Soemarwoto, O. 1980. *Arch. Environ. Contam. Toxicol.,* 9: 437–449. (Cited from Asami, 1991).

Tadano, T., Yonebayashi, K. and Saito, N. 1992. Effect of phenolic acids on the growth and occurrence of sterility in crop plants. In Kyuma, K., Vijarnsorn, P. and Zakaria, A. (Eds.) *Coastal Lowland Ecosystems in Southern Thailand and Malaysia,* pp.358–369, Kyoto Univ.

Tanaka, A. and Yoshida, S. 1970. *Nutritional Disorders of the Rice Plant in Asia.* IRRI, Tech. Bull. No. 10. IRRI, Manila, Philippines.

Van Alphen, J.G. 1975. Salt affected soils in Peru. *ILRI Annual Report,* pp. 7–13. (Cited from Bhumbla and Abrol, 1978).

Van Breemen, N. and Pons, L.J. 1978. Acid sulfate soils and rice. In IRRI (Ed.) *Soils and Rice,* pp. 739–761. IRRI, Los Baños, Philippines.

Van Breemen, N. and Wielemaker, W.G. 1974. Buffer intensities and equilibrium pH of minerals and soils. II. Theoretical and actual pH of minerals and soils. *Soil Sci. Soc. Amer. Proc.,* 38: 61–66.

Van Breemen, N. 1975. Acidification and deacidification of coastal plain soils as a result of periodic flooding. *Soil Sci. Soc. Amer.Proc.,*39: 1153–1157.

Yamada, I. 1986. Vegetation in swampy lowlands in Southeast Asia. In Tropical Agriculture Research Center (K. Kyuma, Ed.) *Swampy Lowlands in Southeast Asia,* pp. 104–196. Norin-Tokei-Kyokai, Tokyo. (In Japanese).

Yoneda, S. 1964. *Pedological and Edaphological Studies on Polder Land*

Soils of Japan. Faculty of Agric., Okayama Univ., Okayama. (In Japanese).

Yoshikawa, T., Jikihara, T., Yoshida, T. and Kusaka, S. 1979. *Res. Rep. Hyogo Agric. Expt. Sta.*, 28: 115–118. (Cited from Jikihara, 1991. In Japanese).

Yoshikawa, T., Motoyama, M., Hashizume, G., Kusaka, S., Jikihara, T. and Yoshida, T. 1978. Distribution of heavy metals in rice grains during the ripening process. *Abstracts of the 1978 Meeting, Soc. Sci. Soil & Manure, Japan*, 24: 155. (Cited from Jikihara, 1991. In Japanese).

Chapter 12: The Paddy Soil/Rice System in the Environment

12.1 Introduction

Paddy rice has been cultivated for thousands of years in Monsoon Asia, enabling the region to support extraordinarily high densities of human populations. This was made possible primarily by the high productivity and stability of rice cultivation. The reasons for this high productivity have been discussed in preceding chapters, particularly in Chapters 8 and 9, in relation to the behavior of individual nutrients in paddy soils. However, it is pertinent to summarize them here, and to further discuss the stability or sustainability aspects of rice cultivation. In this discussion, it is more appropriate to deal with aquatic rice and paddy soils in an integrated system, that is, a paddy soil/rice system.

12.2 Intrinsic merits of the paddy soil/rice system

The paddy soil/rice system has several intrinsic advantages, such as a high natural supply of nutrients, relative indifference to soil tilth, and others. They are elaborated below.

12.2.1 Natural supply of nutrients

Perhaps the most unique feature of the paddy soil/rice system is its high self- or natural-supply capacity for plant nutrients. Basic cations, like K, Ca, and Mg, and readily soluble silica, are supplied by irrigation water. Table 12.1 provides the mean river water quality data for Japanese and Thai rivers (Kobayashi, 1958, 1961). Even with somewhat dilute Japanese river water, the supply of Ca and Mg by irrigation water exceeds crop requirement by a few to several times, and the supply of K and silica satisfies a substantial part of the need of a crop even at a relatively high yield level (see Chapter 9). The higher elemental content of Thai rivers, together with their relatively high sediment load, supplies cationic nutrients and silica in sufficient amounts to produce a crop of rice at a yield level of less than 3 t ha^{-1}.

Table 12.1 Water quality of Japanese and Thai rivers

	Japanese rivers (mean: mg kg⁻¹)	Thai rivers (mean: mg kg⁻¹)
Calcium (Ca)	8.8	19.8
Magnesium (Mg)	1.9	3.7
Potassium (K)	1.2	2.5
Silica (SiO$_2$)	19.0	16.0
Total soluble salts	74.8	115.2
Suspended sediments	29.2	112.0

(Source: Kobayashi, 1958, 1961)

Unless water is polluted, N is not supplied in a meaningful amount by irrigation water, but is supplied by decomposition of soil organic matter, the level of which is kept higher in paddy soils that undergo submergence than in upland soils (see Chapter 7). In addition, it is also supplied by biological fixation to the extent of 30 to 40 kg ha⁻¹ per crop, an amount sufficient to produce 1.5 to 2 t ha⁻¹ of paddy and surpassing that fixed by upland cereal crops by two to three times. Soil P is made available to rice much more readily under submergence than under upland conditions due to the reduction of iron phosphates, and the higher solubility of iron and aluminum phosphates at higher pH under reduced paddy soil conditions.

These self-supplying mechanisms of nutrients help to maintain high yields compared with upland cereals. Kawasaki's (1953) summary, as shown in Table 8.3, indicates that No-P and No-K plots give only a 5% and 4% decrease, respectively, in rice yield compared to the complete fertilizer plot in field experiments. Whereas they respectively show a 31% and 22% decline in yields of upland cereals, including wheat and barley. Even a No-fertilizer plot of rice gives 78% of the complete plot yield, while that of upland cereals gives only 38%. The cause for such a conspicuous difference between rice and upland cereals in their response to fertilizer application may be attributed to the aforementioned nutrient self-supply capacity of paddy soils. This is an important aspect of rice soils, particularly in some developing countries where fertilizers are not used in adequate amounts or are used improperly. Rice production in such conditions is still feasible, and, consequently, the process of 'nutrient mining'—the nutrient depletion of upland cereal crops by continuous cultivation—is less frequent in rice systems.

12.2.2 Detoxification of excessive nutrients

Nitrogen and phosphorus are the two major elements that cause eutrophication of inland water bodies, and today agricultural land is often considered to be an important non-point source of these pollutants, because farmers tend to apply excessive amounts of N and P to obtain higher yields and better quality produce.

However, paddy soils have a built-in mechanism for removing and detoxifying excessive N during the cropping period. Ammoniacal N released from decomposing organic matter or applied as fertilizer is stable in reduced soils and not readily leached out. However, once a thin oxidized layer is differentiated on top of the strongly reduced plow layer under submergence, this layer becomes the site for N transformation. Ammoniacal N is oxidized to nitrate N, which, being an anion, readily moves into the reduced plow layer to be lost to the atmosphere by denitrification. This is in sharp contrast to upland fields that tend to lose much of their applied N as nitrate that pollutes ground and surface water.

Phosphorus is retained by soil particles, particularly in oxidized subsoil, as nearly insoluble phosphates and, thus, it is not readily leached out.

12.2.3 Detoxification of agrochemicals

Although many of the chlorinated hydrocarbons, such as lindane (BHC), dichlorodiphenyltrichloroethane (DDT), and pentachlorophenol (PCP), are today banned for use in agriculture in many countries, it is an established fact that they decompose much faster in paddy soils under submergence than in upland soils (Sethunathan and Siddaramappa, 1978). This may be due to the dechlorination and dehydrochlorination reactions coupled with soil reduction. Diazinon and parathion are deactivated by hydrolysis. Also, some other agrochemicals, such as nitrofen (NIP) and chlornitrofen (CNP), are rapidly deactivated in paddy soils because of the reductive transformation of the nitro-group to the amino-group.

It is fair to mention that some agrochemicals, such as fenthion (MPP) and fenobucarb (BPMC), have a longer half-life in paddy soils than in upland soils. However, as shown in Table 12.2, for many of the commonly used agrochemicals, such as dimethoate, diazinon, fenitrothion (MEP) and chlorothalonil (TPN), the half-life period is shorter in paddy soils than in upland soils by half or less (Kanazawa, 1992).

12.2.4 Relative indifference to soil tilth

It is imperative for upland soils to maintain good soil tilth, because water relations, such as permeability and water-holding capacity, are often more critical than soil fertility for growing good crops. Elaborate management to maintain good structure and consistency of the soil is essential for upland farming and, for this purpose, organic manures have to be applied

Table 12.2 Comparison of half-life period of various agrochemicals between upland and paddy field condition (in days)

Agrochemicals	Upland soil		Paddy soil	
	non-volcanic	volcanic	non-volcanic	volcanic
Diazinon	7–14	7–14	3–7	3–7
Dimethoate	30–60	30–60	14–30	7–30
Disulfoton	30–45	45–60	30–45	60–90
Vamidothion	1–2	2–3	2–3	2–3
Malathion	0–1	0–1	0–2	0–2
MEP	13–29	13–16	7–14	7–14
MPP	6–13	13–20	13–20	30–45
CYAP	3–7	3–7	3–7	3–7
PAP	0–1	0–1	3–7	0–1
NAC	14–21	14–21	14–21	14–21
BPMC	7–14	7–14	56–80	59–114
Cartap	0–3	0–3	0–3	0–3
Methomyl	0–16	0–16	1–3	3–5
TPN	3–7	3–7	0–1	0–1
Trifluralin	30–50	30–50	10–15	10–15

(Source: Kanazawa, 1992)

to activate soil biota besides chemical fertilizers. In contrast, for paddy soils the considerations on tilth are relatively unimportant as long as water for submergence can be secured. In a sense, in paddy soils the structure is intentionally destroyed by puddling to control the excessive downward percolation of water and to provide a better rooting bed for young transplanted rice seedlings. Furthermore, in very clayey soils that quite frequently occur in deltaic sediments, plowing is easier under submerged conditions than under dry upland conditions. However, this is relative. Even in paddy soils, good tilth is desirable when a very high yield, of say more than 10t ha^{-1}, is desired. To control the nutrient status of the soil, water must be managed carefully and for this to be made possible, good soil tilth must be provided and maintained. Some innovative paddy farmers in Japan attain such high yields of rice by the annual application of large amounts of farmyard manure to make deliberate water management possible.

12.2.5 Resistance to soil erosion

Monsoon Asia is one of the most vulnerable regions of the world for soil erosion and land degradation. As discussed in Chapter 2, intensive orogenic movement and active volcanism are the two basic geological features of the region. These features, combined with monsoon rains,

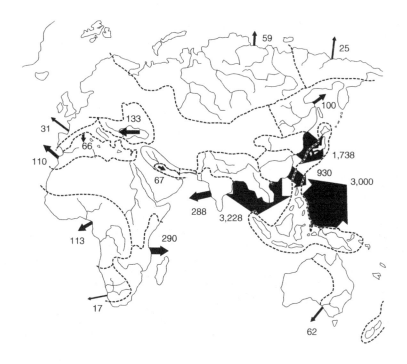

Figure 12.1
Annual discharge of suspended sediments from various drainage basins of the world
(after Milliman and Meade, 1983) (Source: Stangel and Uexkuhl, 1989)

characterised by high intensity storms, result in an exceptionally large amount of sediments that is deposited in the lowlands or carried into the sea. Figure 12.1 indicates the high sediment discharge in Monsoon Asia compared with other continents. Rice cultivation is an adaptation to this feature, in that it has made possible an efficient use of the extensive lowlands in Monsoon Asia and provided an effective measure for conserving the region's land and soil.

Each paddy field is leveled and enclosed with a bund to retain irrigation water. Even on sloping lands elaborate terraces are constructed for paddy fields. This is a perfect conservation and protection measure against soil erosion and probably the most important advantage of the paddy soil/rice system over upland farming systems. The sustainability of paddy rice cultivation is largely based on this resistance to soil erosion. During periods of flooding, water flows over the bunds producing some soil loss. In general, however, this is infrequent.

12.2.6 Tolerance to monoculture

Rice is cropped every year without rotation yet no particular harm results unlike the monoculture of upland crops, for which various biological and non-biological problems, collectively called 'soil sickness', inevitably occur and the adoption of crop rotation is demanded.

Annually alternating aerobic and anaerobic conditions under rice cropping is an effective control measure for those pests and diseases that cause soil sickness. No living creature can survive both the strongly aerobic and strongly anaerobic conditions that paddy soils experience. Moreover, the nutrient self-supply mechanisms and relative indifference to soil tilth allows rice to tolerate non-biological problems that emerge from monoculture situations. In fact, the conversion of upland crops to paddy rice is practiced routinely in Japan specifically to avoid these problems resulting from continuous upland cropping.

12.2.7 Relative ease of weeding

Weeds are a serious problem in upland farming in humid Monsoon Asia. If proper weeding were not practiced, weeds would proliferate and negate all the farmers' efforts. The paddy soil/rice system has a relative advantage over upland farming in controlling weeds. Under submerged paddy field conditions, the biomass of weeds decreases to one-third that of the water-saturated upland, and even to one-sixth of that of ordinary upland conditions. Moreover, submergence selectively eliminates terrestrial C_4[1] weeds, such as cogon grass (*Imperata cylindrica* (L.)P. Beauv.), green foxtail (*Setaria viridis* (L.)P.Beauv.) and crabgrass (*Digitaria adscendens* (H.B.K.) Henr.), that are most competitive with rice, which is a C_3[1] plant (Yoshida, 1982). Of course, aquatic weeds are equally troublesome to farmers, but they are not so numerous and uprooting is relatively easy under water. In some fallowed paddy fields in Japan, water is deliberately kept to suppress terrestrial weeds.

12.2.8 Carbon sequestration

As already discussed in Chapter 7 and elsewhere, organic matter tends to accumulate more in paddy soils than in upland soils even under the same climatic conditions. In the face of global warming, this fact may be viewed from the standpoint of efficiency in carbon sequestration and counted as another intrinsic merit of paddy soils compared with upland soils.

In Japan, the organic carbon content of soils is sometimes more than doubled by their long-term use for paddy cultivation, compared with the same soil cultivated for upland crops (Mitsuchi, 1974). As many management and biophysical factors could be involved in the accumulation of carbon in paddy soils, it is difficult to generalize the Japanese experience. However, the very fact that organic matter tends to be more efficiently preserved when submerged helps paddy soils sequester more carbon than upland soils.

12.3 Intrinsic demerits of the paddy soil/rice system

The high sustainability of a system is a necessary condition for conserving the environment, of which the system is a part. But even such a system can harm the environment. The paddy soil/rice system has certain intrinsic features that adversely affect the environment through air and water pollution.

12.3.1 Air pollution

As already discussed, under submerged conditions various chemical transformations proceed successively: oxygen depletion, denitrification, reduction of manganic and ferric oxides, fermentation of organic substrates producing various organic compounds, sulfate reduction and methane fermentation. Of these, denitrification, sulfate reduction, and methane fermentation emit gases that are detrimental to the environment.

1) Nitrous oxide and sulfur compounds

Denitrification emits not only dinitrogen gas but also nitrous oxide gas (N_2O), although the factors that regulate the relative proportion of nitrous oxide in the total gas emissions are not clearly understood. Nitrous oxide is a greenhouse gas that is 200–300 times more efficient than CO_2, and it also plays a role in the catalytic decomposition of the ozone in the stratosphere. However, it is also emitted by the oxidative nitrification process, and this process has become more important in the overall global emission of N_2O. Thus, denitrification in paddy soils is currently regarded as rather minor as a source of nitrous oxide. As a matter of fact, there are studies that have proved the negligible emission of nitrous oxide from paddy soils (DeDatta and Buresh, 1986; Bronson and Singh, 1995). Furthermore, Minami and Fukushi(1987) reported that paddy soils may act as a sink for N_2O under strongly reduced conditions.

Sulfate reduction emits hydrogen sulfide in strongly reduced conditions. However, it is mostly precipitated as ferrous sulfide in paddy soils and is rarely released into the atmosphere (see Chapter 9). Recent studies on sulfur-containing gases in paddy soils revealed that dimethyl sulfide [$(CH_3)_2S$], carbonyl sulfide (COS), and carbon disulfide (CS_2) are more important than hydrogen sulfide in quantity. However, even these compounds are emitted in very small quantities and their impact on the atmospheric environment, for example as a source of acid rain, may be negligible (Kanda, 1994).

2) Methane

Methane is produced under extremely reductive conditions in the presence of readily decomposable organic matter. It is an effective greenhouse gas (about 20–60 times more efficient than CO_2) and recently its atmospheric concentration has been increasing by 1% annually. Paddy soils are considered to be an important source of methane. At present, the contribution of paddy soils to annual global methane emissions is estimated at 11% or 60 Tg, although the estimates range widely from 20 to 150 Tg (IPCC, 1995). Thus, methods for suppressing methane emissions from paddy soils, are of great concern. Many studies have been and are still being conducted to determine ways and means for reducing methane emissions from paddy fields.

As described in Chapter 4, methane is produced microbiologically under strongly reductive submerged conditions in two ways—the reduction of carbon dioxide with hydrogen gas and decarboxylation of acetic acid abundant in submerged paddy soils:

$$4H_2 + HCO_3^- + H^+ = CH_4 + 3H_2O \ (CO_2 \ reduction)$$
$$CH_3COO^- + H_2O = CH_4 + HCO_3^- \ (decarboxylation)$$

The factors influencing methane emissions from paddy fields are (Yagi, 1994):

- the redox potential (below -200 mV);
- the ratio of oxidizing capacity ($O_2 + NO_3^-$ + readily reducible $Mn^{III, IV}$ and Fe^{III}) to reducing capacity (readily decomposable organic matter)—this is proportional to the ratio of CO_2/CH_4;
- the quantity and quality of organic matter;
- the temperature ($20 < 40$ [optimal] $<60°C$);
- competition with sulfate reduction (reductive methane oxidation);
- others (such as soil pH, activity of methane oxidizing bacteria, biomass of rice, fertilizer application, salt concentration).

Of these factors the 'quantity and quality of organic matter' may be the easiest to manage. In practice, the use of farmyard manure made from rice straw greatly reduces methane emissions compared with its direct application.

Special attention should be paid to the 'biomass of rice' factor. As rice grows more vigorously, more methane is emitted, and this may be due to a greater supply of readily decomposable organic matter in the form of root exudates, dead roots and leaf sheaths. Thus, it appears that efforts to increase rice production would, to some extent, also increase methane emissions.

It is also noteworthy that the activity of methane-oxidizing bacteria in the thin surface layer of wetland soils may be much higher than that of methane-forming bacteria in the reduced subsurface layer by a factor greater than 10. Therefore, unless there are aquatic plants like rice and reeds that can transfer methane directly to the atmosphere through their stalks, little methane would be emitted from the soil surface. More than 90% of the methane emitted from paddy fields is transmitted passively via rice stalks.

The factors and processes relevant to methane production in, and its release from, paddy soils are illustrated in Figure 12.2. Using an example

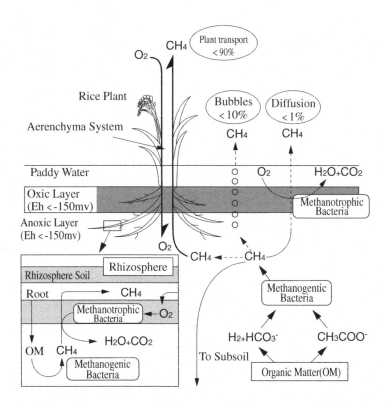

Figure 12.2
Production, oxidation and emission of methane in paddy soils (Source: Yagi, 1997)

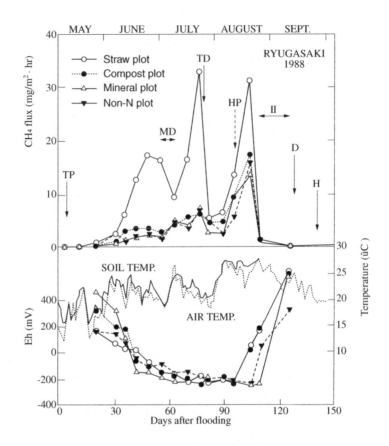

Figure 12.3
Seasonal variation of methane flux as affected by soil conditions and management
practices in a gley soil at Ryugasaki, Ibaraki Pref. (TP: tramsplanting, MD: mid-
term drying, TD: top-dressing, HP: heading period, II: intermittent irrigation, D:
drainage, H: harvest (Source: Yagi, 1997, partly modified)

from Japan, Figure 12.3 shows the changes in the methane flux over the
course of a rice crop as affected by various management factors.

Measures to control methane emissions from paddy soils may be put into
two categories: input management and water management. Among the
inputs, readily decomposable organic matter should be avoided as much
as possible. It should be applied after conversion to farmyard manure. Water
management to control methane emission is discussed later.

12.3.2 Water pollution

Despite the detoxification mechanisms of paddy soils, N and P can enter drainage ditches and move into rivers, ponds and lakes. In Japan, this type of water pollution occurs mainly during the period of puddling and transplanting, as seen in Figure 12.4 (Tabuchi, 1976). This is caused by draining the surface water immediately after puddling in order to transplant young small seedlings.

One estimate of water pollution loads from paddy soils is given at 10–46 kg ha^{-1} of total N and 0.5–9 kg ha^{-1} of total P (Kunimatsu, 1989,1995). However, if we take a more general look, about 50% of paddy soils are sources of water pollution, while the remaining 50% are purifying water. Where irrigation water is cleaner or a high dose of N fertilizer is used, paddy

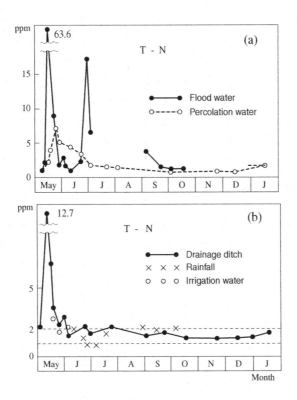

Figure 12.4
(a) T-N concentration in a paddy soil, (b) T-N concentration in drainage ditch, rainfall and irrigation water (Source: Tabuchi, 1976)

soils become a source of pollutants, whereas when irrigation water has more than 2 mg litre^{-1} of total N they act as a sink (Morikawa, 1987).

Besides nutrient pollution, surface water drained immediately after puddling also causes muddy water due to dispersed clay. A large amount of the sediment suspended in river water, often measured during the puddling period, also carries a significant amounts of nutrients that pollute or eutrophicate water bodies.

12.3.3 High water use

Water is an integral part of the paddy soil/rice system and a very high quantity of water is a prerequisite for the system. Normally, some 1.5 m of water is used for one crop of rice. Of course, the water in paddy fields is not only used repeatedly, flowing from one field to another, but it also recharges ground-water. However, with today's very competitive water allocation among the urban, industrial, and agricultural sectors, the availability of water for rice cultivation is being gradually restricted. In Japan, irrigation systems have been developed throughout history and along with them a very strong 'water rights' have been established to protect rice growers. Even with this traditional right water is becoming increasingly difficult to secure for agriculture as the urban population increases and industries flourish. Thus, it is imperative to restrict excessive water use by the paddy soil/rice system.

12.4 Management factors to attain sustainability

As reviewed above, there are merits and constraints that are intrinsic to the paddy soil/rice system. Management factors can augment or lessen these merits and constraints. What, then, are the management practices used by more progressive growers today? The following examples are taken mainly from Japan.

12.4.1 Fertilization

As stated earlier, unless an excessive amount of nutrients is applied, many paddy soils function as an N sink when growing rice, except in the period of puddling and transplanting. In order to augment this, it is necessary to refrain from an excessive use of fertilizers and prevent the draining of irrigation water immediately after puddling.

To avoid an excessive use of fertilizers, it is necessary to raise the use efficiency of fertilizers by plants. A transplanter, equipped with a fertilizer applicator, was developed in Japan around 1980 to help do this. This can place basal doses of fertilizers in the vicinity of rice roots while transplanting the rice seedlings. With the use of this transplanter-cum-fertilizer applicator, not only is the use efficiency of fertilizers greatly improved, but also the nutrient outflow due to puddling (except that adsorbed on soil particles) is almost totally arrested.

More recently, the use of coated urea as the basal dose in the no-till or no-puddle/transplanting system has been introduced. In this system, puddling is not practiced before transplanting. Also, the basal dose of N is applied as a controlled-release type fertilizer and N is released gradually as the rice grows. Top-dressings of N may or may not be applied depending on the performance of the rice.

To go one step further, all of the N necessary for one crop of rice may be applied to the nursery boxes as coated urea before seeding. This type of coated urea can release N in a sigmoidal pattern over 100 days or even longer, depending on the type of fertilizer, so that top-dressing may not be required, enabling considerable saving of labor. The rice seedlings are machine transplanted with fertilizer capsules enmeshed with their roots. The use efficiency of N fertilizer applied in this manner is very high as shown in Table 12.3 (Kaneta *et al.*, 1994). It is therefore possible to reduce N fertilizer by 30–40% of the ordinary dose. The yield of rice compares with that from ordinary practice, and the high cost of fertilizer may well be compensated by the savings in labor and fertilizer use.

12.4.2 Water management

In modern paddy rice cultivation, water should be managed to achieve multiple goals: water conservation; minimizing water pollution; and suppressing the emission of air pollutants, particularly methane.

Recycling irrigation water meets the first two goals. Drainage water from paddy fields is pooled before being discharged to inland water bodies and

Table 12.3 Use efficiency of basal nitrogen at the maturation stage (cultivar: Akita-Komachi)

Year	Type of fertilizer	Place of application	Rate of utilization (%)
1990	Ammonium sulfate	Soil surface	9.3
	ditto	Vicinity of roots	32.5
1991	Coated urea	Soil surface	60.5
	ditto	Vicinity of roots	78.9
	ditto	Nursery box	83.2

(Source: Kaneta *et al.*,1994)

pumped upstream of irrigation channels to be reused for irrigating paddy fields. The relatively high cost of installation of this system is one problem and the reluctance of farmers to use recycled water is another, particularly in water-rich areas such as those adjacent to Lake Biwa in Shiga Prefecture.

Mid-term drying, at the end of the effective tillering stage, is practiced widely in Japan to suppress further tillering, and supply oxygen to the root zone to restore the vigor of the rice roots. Subsequently, intermittent irrigation may be adopted to maintain the soil bearing strength until harvest, except during the heading period. In intermittent irrigation, water is introduced, say, for three or four days and then drained for five days, then the pattern is repeated. This is an appropriate water conservation practice for a water-poor area, where irrigation water is supplied alternately among sub-areas.

Water management practices like mid-term drying and intermittent irrigation should also be evaluated as a means for suppressing methane emissions from paddy soils. According to Yagi (1994), total methane emissions from paddy fields for one rice-cropping season was greatly reduced by practicing one or two short mid-term dryings. However, such practices are possible only where perfect irrigation and drainage systems exist.

12.4.3 Low-input rice cultivation—an alternative technology

By and large current rice cultivation in Japan is based on high-input and high-output technologies. A widely used rice management system in Japan consists of dense planting and a high N dose to promote the vigorous early growth considered essential to obtain high yields. However, according to Hashikawa (1985) rice grown by this system tends to have a large number of sterile and less productive tillers and, consequently, produces less grain than expected from its early growth. This is particularly true in warmer southwestern Japan.

As an alternative to the conventional system, Hashikawa proposes a system that achieves 'autumn vigor', which is opposite to 'autumnal decline' or the *akiochi* growth pattern, so growth is slower at first, but increases later. His major recommendations include:

- enhancing the potential tillering ability of seedlings;
- setting the planting density as low as possible, especially by reducing the plant number per hill, to the extent that an adequate crop biomass can be secured;
- reducing the total amount of fertilizer N, by minimizing the basal dose and supplementing it with top-dressings from the late vegetative stage onward at a rate that meets the crop requirement.

The result of Hashikawa's four year experiment in Shiga Prefecture gave consistently higher yields (5.6 to 6.3 t ha^{-1} as brown rice) than the average for the locality, using only half the conventional amount of N (50 kg N ha^{-1} as coated urea) and half the planting density (44 plants m^{-2}) (Hashikawa, 1996).

This is just an example, but it suggests that there is still some room for moving towards low-input technology, while maintaining a fairly high productivity. It is hoped that concerted efforts among agronomists, plant protection specialists, and soil scientists will bring about many more examples of low-input, environmentally sound rice cultivation technologies in the near future.

12.5 The multi-functionality of the paddy soil/rice system

Thus far, issues directly concerned with the sustainability of the paddy soil/ rice system have been discussed. However, the system has many more implications for the environment and society. Some of these points are discussed next to emphasize the importance of the paddy soil/rice system in preserving the environment of Monsoon Asia.

Paddy fields can retain huge volumes of water during the rainy season. If the bund height is assumed to be 20 cm, 1000 ha of paddy field can retain 2 Mt of water. In Japan, one such calculation revealed that paddy fields were playing a far greater role in storing water than all of the nation's dams. This capacity can be used to adjust the rate of surface run-off and river flow at times of heavy rains, thus preventing floods downstream. Also, the water used for submerging paddy fields slowly recharges the ground-water, which in turn slowly drains into rivers to maintain their unceasing flow.

Disruption of this function could cause serious disasters. In Japan, many cities have lost paddy fields due to urbanization, and as a result, the frequency of flooding has increased. Being cognizant of this cause-effect relationship, some cities are now subsidizing owners to maintain paddy fields in their suburban areas, regardless of whether the fields are planted with rice or just left fallow.

The passage of polluted irrigation water through paddy fields may reduce the level of pollutants. A notable example of this in Japan is the use of water flowing from tea plantations on terraces to irrigate paddy fields on low-lying floodplains, so making use of the natural sequence of landforms. The surface run-off and ground-water from tea plantations are notoriously high in N content, often >30 mg NO$_3$-N litre^{-1}, which poses a serious disposal problem in many tea-growing areas. Nitrogen in the water introduced to lower areas from tea plantations is partly utilized by rice and largely denitrified in reduced paddy soils. Thus, paddy fields have the dual roles of resource utilization and the safe disposal of pollutants.

There are other functions. Paddy fields with submergence water favorably modify the local climates. They also provide habitat for many fish, aquatic insects, and animals in their natural state, contributing to the enhancement of biodiversity. However, most importantly, rice cultivation, along with its environment, is fundamental to the culture of the people of Monsoon Asia. Many customs and rituals are associated with rice cultivation. Human relations and ideologies, or value systems, have been developed over hundreds or thousands of years of paddy rice farming. Even the landscapes in Monsoon Asia—typically represented by terraced paddy fields like those in Banaue, of northern Luzon in the Philippines—have been constructed for rice cultivation. The concept of cherishing the cultural heritage associated with the paddy soil/rice system is truly the basis for the preservation of nature and the environment of Monsoon Asia.

12.6 Future prospects for the paddy soil/rice system

12.6.1 Intensification of rice cultivation

As discussed in Chapter 2, Monsoon Asia is facing an acute land shortage and the only way to feed the region's ever-increasing human population is to intensify its agriculture, both on paddy lands and uplands. In fact, the area of cultivated uplands in Monsoon Asia totals 290 M ha, which is more than twice that of paddy lands (136.5 M ha). Therefore, it is vital to intensify not only paddy cultivation, but also crop cultivation in the uplands.

Previous experience, however, tells us that any attempt to intensify upland cultivation often leads to land degradation through nutrient mining and soil erosion. Loess-derived soils in the north are notoriously susceptible to erosion and widespread ultic soils (Ultisols, ultic Inceptisols, or acrisols) in the south, including the whole of tropical Asia, are both infertile and highly erodible. Effective soil conservation measures as well as integrated nutrient and pest management are essential to successfully intensify food production on upland soils in Monsoon Asia. Also, good care must be taken not to pollute the environment as more chemical fertilizers and synthetic pesticides are used in intensive crop cultivation.

In contrast, it is much safer to intensify paddy rice cultivation. As already mentioned, there are built-in mechanisms operating in the paddy soil/rice system to counter soil erosion, to gain a natural supply of nutrients, and to detoxify excessive nutrients and pesticides. However, all these mechanisms will only work more effectively if irrigation and drainage can be controlled. In the greater part of tropical Asia, the infrastructure for irrigation and drainage has not been developed to the level required by intensive rice cultivation. Table 12.4 shows the distribution of different rice ecosystems in both continental and insular Southeast Asia, (IRRI, 2002). Insular Southeast Asia had 57% of its paddy land area under reasonably good, but still far from perfect, irrigation systems, whereas only 31% of rice in continental Southeast Asia is irrigated and, nearly 5% is under deep water. South Asia has improved its irrigation systems greatly over the last 30 years, mainly by means of tube wells, but irrigated rice land is still a little less than a half the total. In this case, the figures in Table12.4 indicate that tropical Asia still has much potential for infrastructure development and intensification of rice cultivation. The large rivers in continental tropical Asia appear to be difficult to control, even with modern engineering skills, and the investment required to install irrigation and drainage facilities is tremendously high. Nevertheless, it should still be possible to realize the potential as demands increase.

Chemical fertilizers and pesticides, as elements of agricultural inten-sification, are already quite widely used in Monsoon Asia, particularly where the infrastructure has been developed. This, along with the adoption of modern high yielding varieties, has brought about a very remarkable increase in the yield of rice in tropical Asia, as shown in Table 12.5. The figures for the 1960s show yields before the green revolution, those for the 1970s indicate the early days of the green revolution, while those for 2000 are indicative of present rice yield in tropical Asian countries.

Remarkable as it is, the mean rice yield of tropical Asia in 2000, 3.23 t ha^{-1}, was still just a little over half the mean yield, 6.22 t ha^{-1}, of China, Korea, and Japan (East Asia). This lower yield is definitely related to the

Table 12.4 Percentage distribution of various rice ecosystems in tropical Asia

Rice ecosystems	Southeast Asia		South Asia
	Continental	Insular	
Irrigated	31.4	57.2	48.9
Rainfed lowland	60.2	33.4	36.4
Deep water	4.6	0.1	4.4
Upland	3.8	9.3	10.3

(Source: IRRI, 2002)

Table 12.5 Changes of rice yield in tropical Asian countries, from early 1960s to 2000

Country	1961–1965 mean, kg ha⁻¹	1972–1974 mean, kg ha⁻¹	2000 kg ha⁻¹
Bangladesh	1679	1774	3348
Cambodia	1077	1217	2009
India	1480	1657	3008
Indonesia	1761	2514	4426
Malaysia	2503	2943	2941
Myanmar	1642	1680	3333
Pakistan	1417	2296	3027
Philippines	1257	1575	3075
Sri Lanka	1914	2552	3177
Thailand	1775	1818	2329
Vietnam	2034	2439	4253

Notes: Figures for the 1960s and 1970s for Malaysia are for West Malaysia, and those for Vietnam are for the then South Vietnam.
(Source: FAO, 1974; FAOSTAT, 2001)

lower average amount of N fertilizers applied per hectare in tropical Asia, which is often less than half that of East Asia. This in turn is related to the general level of infrastructure development among the other things.

12.6.2 Research needs for the future

Whenever a new concept or methodology evolves, either within soil science itself or in other disciplines, it is necessary for soil scientists to test it, adapt it, or modify it in relation to their own research area. It is not possible to list all the necessary tasks to be tackled, but the author is of the view that the following two topics are the most urgent, one being of particular importance to the tropics, while the other is of more general relevance.

1) Long-term sustenance of high rice yields in the tropics

For more than a decade, a major concern in tropical Asia has been that increases in rice productivity hit a ceiling and no break-through to raise yields has yet emerged. Even worse, a high rice yield (5–6 t ha⁻¹) once attained could not be sustained on the same field for even 10 years in many places, despite the addition of sufficient fertilizers, pesticides, and carefully managed irrigation (Greenland, 1997). One example is reproduced from Cassman and Pingali (1995) in Figure 12.5. Although the data is taken from an exceptionally intensively managed plot with three crops per year,

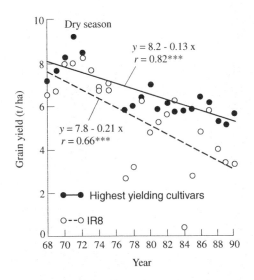

Figure12.5
*Dry season yield trends for the highest yielding variety and IR-8 for plots receiving
150 kg of fertilizer N per crop, in the three crops per year long-term continuous
cropping experiment at IRRI (Source: Cassman and Pingali, 1995)*

the decline in the yield of the varieties studied is too abrupt for the
elaborate management presumably given to the plot. The most serious fact
to be read from the figure is that this result occurred at an IRRI farm, where
all the necessary hardware and software to support rice cultivation should
have been available.

In temperate countries like Japan and China, yields of 6 t ha^{-1} or more
have been sustained for more than 20 years (Greenland, 1997). No such a
long-term steady decline has been reported for similarly high yields in
Australia or the USA. Apparently, the decline in rice yields appears to be
directly associated with biophysical conditions in the tropics. Elucidation
of the relevant factors is of paramount importance for the future of rice
cultivation in tropical Asia, and the role of soil scientists in this task is
paramount.

2) Precision agriculture for paddy rice cultivation

For economic and environmental reasons, site-specific management of
agriculture (or precision agriculture) is advocated and studied for many
types of cultivation. In Japan, pioneering work on precision agriculture is
being conducted on paddy lands.

In addition to the intrinsic heterogeneity of paddy fields (Yanai et al., 2000), recent land consolidation projects in Japan tend to enhance their heterogeneity. Land consolidation is being promoted in paddy areas to construct large fields from finely-divided lots and facilitate mechanized farming. The surface soil is removed and later replaced after the land is levelled. However, the new surface soil becomes quite heterogeneous, partly because the former surface soils had a different management history and partly because the admixture of the subsoil. Therefore, managing soil fertility in the new field is often very difficult. Some parts of the field may suffer an N shortage, while other parts have an excess, resulting in lodging. Thus, there is a practical need for site-specific management.

Toriyama et al. (1999, 2002) conducted basic research to develop the necessary research methodology and instrumentation for producing a high and uniform rice yield in consolidated paddy fields. They used a field of about 1 ha for the experiment and divided it into 72 grids of 12.5 × 10m and installed a mapping system equipped with a GPS (Global Positioning System) and a digital camera to monitor crop growth (plant cover ratio) on a tractor. They also attached a specially-developed mechanized soil auger to enable repeated sampling of each grid to a depth of 30 cm. Available soil N was analylsed and correlated with crop uptake and growth. In the course of the research, various sensing methods were also developed; near-infrared spectrophotometric sensing of total N in the plow layer and subsoil (Toriyama et al., 2000), and near-infrared real-time sensing of soil organic matter and electric conductance (Shibusawa et al., 1999).

The results gained by Toriyama et al. revealed that the brown rice yield in the experimental field varied widely between 341–601 g m^{-2}, but it was highly positively correlated with the N uptake of rice plants. They considered that the difference in N uptake was caused by diffrneces in the amount of N mineralized from soil organic matter. Thus, using the site-specific information of available soil N, it was possible to adjust the N fertilizer dose across the field to secure a higher and more uniform rice yield. They are also trying to monitor and map rice growth at different stages so that they can make necessary site-specific amendment to the crop or soil during the cropping season. Such elaborate management is also useful for producing high quality rice by controlling the protein content of rice grains. As a by-product of this research, they found that the high heterogeneity of subsoil fertility affects the performance of rice very significantly (Toriyama, 2001).

The environmental significance of precision agriculture is obvious. In intensive rice farming, chemical fertilizers and synthetic pesticides are applied uniformly and often excessively. Site-specific management can lessen the danger of excessive chemical use and significantly reduce the risk of environmental pollution. However, precision agriculture is still being developed. Many kinds of sensors need to be developed to enable real-time monitoring of the physical and chemical condition of soils as well as the physiological status of rice to detect incidences of pests and diseases. Interdisciplinary cooperation among soil scientists, crop scientists and machinery specialists will be imperative if precision agriculture is to be made a really useful tool for intensive rice cultivation.

Notes

(1) C_3 and C_4 are two types of carbon assimilation pathways found in photosynthesis. C_4 plants are more efficient in photosynthesis than C_3 plants, particularly in strong sun light.

References

Bronson, K.F. and Singh, U. 1995. Nitrous oxide emissions from flooded rice. In Peng, S., Ingram, K.T., Neue, H.-U., and Ziska, L.H. (Eds.) *Climate Change and Rice*, pp. 116–121, IRRI, Springer Verlag.

Cassman, K.G. and Pingali, P.L. 1995. Extrapolating trends from long-term experiments to farmers' fields: the case of irrigated rice systems in Asia. In Barnet, V., Payne, R. and Steiner, R. (Eds.) *Agricultural Sustainability: Economic, Environmental and Statistical Considerations*, pp. 63–84. J. Wiley, Chichester, U.K. (Cited from Greenland, 1997).

De Datta, S.K. and Buresh, R.J. 1986. Integrated nitrogen management in irrigated rice. *Adv. Soil Sci.*, 10: 143–169.

Food and Agriculture Organization 1974. *Production Yearbook*. Vol. 28, FAO, Rome.

Food and Agriculture Organization 2001. FAOSTAT. FAO, Rome.

Greenland, D.J. 1997. *Sustainability of Rice Farming*. CAB Int'l, U.K. and IRRI, Los Baños.

Hashikawa, U. 1985. *Principles of Rice Crop Management*. Nobun-kyo, Tokyo. (In Japanese).

Hashikawa, U. 1996. Possibility of stable rice production with lower-input. In Hashikawa, U. (Ed.) *Possibility of Low-Input Rice Cultivation*. Fumin-Kyokai, Osaka. pp. 13–234. (In Japanese).

Intergovernmental Panel on Climate Change (IPCC) 1995. *Climate Change 1994, Radiative Forcing of Climate Change and an Evaluation of the IPCC 1992 IS92 Emission Scenarios.* Cambridge Univ. Press, New York.

International Rice Research Institute 2002. *World Rice Statistics, 2002.* IRRI, Los Baños, Philippines.

Kanazawa, J. 1992 *Environmental Science of Agrochemicals.* Godo-Shuppan, Tokyo. (In Japanese).

Kanda, K. 1994. Sulfur-containing gases. In Minami, K. (Ed.) *Pedosphere and Atmosphere—Gas Metabolism of Soil Ecosystem and the Global Environmen—*, pp.107–121. Asakura-Shoten, Tokyo. (In Japanese).

Kaneta, Y., Awasaki, H. and Murai, Y. 1994. The non-tillage rice culture by single application of fertilizer in a nursery box with controlled-release fertilizer. *Jap. J. Soil Sci. Plant Nutr.,* 65: 385–391. (In Japanese).

Kawasaki, I. 1953. *Natural Supply of Three Major Elements from Soils of the Major Arable Land in Japa*n. Nippon Nogyo Kenkyusho, Tokyo. (In Japanese).

Kobayashi, J. 1958. Chemical studies on the quality of river water in the countries of southeast Asia: Quality of water in Thailand. *Nogaku-Kenkyu,* 46: 63–112. (In Japanese).

Kobayashi, J. 1961. Mean water quality of the Japanese rivers and its characteristics. *Nogaku-Kenkyu,* 48: 63–106. (In Japanese).

Kunimatsu, T. 1989. Reduction of pollution load. In Kunimatsu, T. and Muraoka, K. (Eds.) *River Water Pollution and Modeling Analysis,* pp.208–226. Gihodo, Tokyo. (In Japanese).

Kunimatsu, T. 1995. Water resources and water environment. In Kyuma, K. and Soda, O. (Eds.) *Agriculture and the Environment,* pp.73–147. Fumin-Kyokai, Osaka. (In Japanese).

Minami, K. and Fukushi, S. 1987. Emission of nitrous oxide (N_2O) from agro-ecosystem. *Japan Agric. Res. Quarterly (JARQ),* 21: 22–27.

Mitsuchi, M. 1974. Characters of humus formed under rice cultivation. *Soil Sci. Plant Nutr.,* 20: 249–259.

Morikawa, M. 1987. Changes in nitrogen and organic matter in irrigation water in paddy fields. *Res.Output*(MAFF Res. Council), 184: 153–155. (In Japanese).

Sethunathan, N. and Siddaramappa, R. 1978. Microbial degradration of pesticides in rice soils. In IRRI (Ed.) *Soils and Rice,* pp. 479–498. IRRI, Los Baños, Philippines.

Shibusawa, S., Hirako, S., Otomo, A., Sakai, K., Sasao, A. and Yamazaki, K. 2000. Real-time soil spectrophotometer for *in-situ* underground sensing. *Jpn. J. Agric. Machin.,* 62: 79–86. (In Japanese).

Stangel, P.J. and Von Uexkull, H.R. 1990. Regional food security:

demographic and geographic implications. In IRRI (Ed.) *Phosphorus Requirements for Sustainable Agriculture in Asia and Oceania*, pp. 21–43. IRRI, Los Baños, Philippines.

Tabuchi, T. 1976. Outflow of fertilizer from the paddy fields. *Soil Phys. Cond. Plant Growth, Japan*. 33: 16–20. (In Japanese).

Toriyama, K. 1999. Recent situation of the rice production and environmentally conscious agriculture in Japan and the demand for the soil science. *Pedologist*, 43:110–116. (In Japanese).

Toriyama, K. 2001. Soil and plant nutrition science now developed from field study—Acquisition and analysis of data from the new view point—: 1. Growth of rice plant in the large size paddy field and the variability in soil nitrogen fertility. *Jpn. J. Soil Sci. Plant Nutr.*, 72: 453–458. (In Japanese).

Toriyama, K., Shibata, Y., Sasaki, R. and Sugimoto, M. 2002. Field trials of a site-specific nitrogen management system for paddy rice in Japan. *Proc. 6th Int'l Conf. Precision Agric. and Other Resource Management, Minneapolis*. (CD-ROM).

Toriyama, K., Tatsubayashi, K., Shibata, Y., Sasaki, R., Kobayashi, K., Chosa, T., Asano, O. and Hirokawa, M. 2000. Rapid and nondestructive determination of total nitrogen in soil profile by NIRS method. *Abstracts of the 2000 Meeting, Jpn. Soc. Soil Sci. Plant Nutr.*, 46: 301 (In Japanese).

Yagi, K. 1994. Methane. In Minami, K. (Ed.) *Pedosphere and Atmosphere— Gas Metabolism of Soil Ecosystem and the Global Environment—*, pp. 55–84. Asakura-Shoten, Tokyo. (In Japanese).

Yagi, K. 1997. Methane emissions from paddy fields. *Bull. Nat. Inst. Agro-Environ. Sci.*, 14: 96–210. (In Japanese).

Yanai, J., Lee, C.K., Umeda, M. and Kosaki, T. 2000. Spatial variability of soil chemical properties in the paddy field. *Soil Sci. Plant Nutr.*, 46: 473–482.

Yoshida, T. 1982. Chapter 7 Health of rice roots and nutrient uptake. In Yamane, I. (Ed.) *Paddy Soil Science*, pp. 280–307. No-Bun-Kyo, Tokyo. (In Japanese)

Index